J. William Langston, M.D.

and Jon Palfreman

The Case of the Frozen Addicts

J. William Langston, M.D., is president of the Parkinson's Institute in Sunnyvale, California, and director of its basic research programs. An internationally recognized authority on Parkinson's disease, Dr. Langston is also the North American editor of the journal *Neurodegeneration,* and chairman of The Parkinson's Epidemiology Research Committee.

Jon Palfreman is an award-winning writer and producer of medical and scientific documentaries. He is a senior producer at WGBH television in Boston.

The Case of the Frozen Addicts

▪ ▪ ▪ ▪ ▪ ▪

J. William Langston, M.D.
and Jon Palfreman

Christof Koch
September 20th 1996
Caltech

VINTAGE BOOKS
A Division of Random House, Inc.
New York

To Connie, George, Juanita, David, Bill, and Toby

FIRST VINTAGE BOOKS EDITION, JULY 1996

 Published in the United States by Vintage Books, a division of Random House, Inc., New York, and simultaneously in Canada by Random House of Canada Limited, Toronto. Originally published in hardcover by Pantheon Books, a division of Random House, Inc., New York, in 1995.

The Library of Congress has cataloged the Pantheon edition as follows:
Langston, J. W. (James William)
The case of the frozen addicts / J. William Langston and Jon Palfreman.
p. cm.
Includes index.
ISBN 0-672-42465-2
1. Parkinsonism—Case studies.
2. Methylphenyltetrahydropyridine—Physiological effect.
3. Designer drugs—Toxicology.
4. Fetal tissues—Transplantation—Case studies.
5. Parkinsonism—Animal models. I. Palfreman, Jon.
II. Title.
[DNLM: 1. Parkinson's disease—therapy—case studies.
2. Fetal tissue transplantation—case studies. 3. Brain tissue transplantation—case studies. 4. Substantia nigra—embryology—case studies. 5. Substantia nigra—transplantation—case studies.
WL359L295c 1995]
RC382.L37 1995
616.8′33—dc20 94-43202
CIP
Vintage ISBN: 0-679-74708-7

Random House Web address: http://www.randomhouse.com/

Printed in the United States of America
10 9 8 7 6 5 4 3 2 1

Contents

Authors' Note

While the story told in this book arose directly from the experiences of one of its authors, Bill Langston, we decided nevertheless to write the book in the third person. This gave us more freedom to handle the complexities of the plot, and to include key events at which Bill Langston was not present.

The events described in this book are based on hundreds of documents gathered over the past thirteen years and on extensive interviews with all the major participants. Where unwitnessed or unrecorded personal exchanges have been cast in dialogue form, the participants have had an opportunity to see the passages to check that they are a fair depiction of what actually occurred. For legal reasons, some of the names of those involved in illegal drug manufacture have been changed.

The Case of the Frozen Addicts

Prologue

San Jose, California, July 1, 1982

George Carillo knew something was wrong the moment he injected the heroin. His arm burned as if hot lead were flowing into his veins, giving him a stunning high, the best he had had for years. Then he began to hallucinate strangely, trying to walk through doors that weren't there, hurting himself each time he plowed into a wall. George vaguely wondered about those four bindles he had bought on the street in Mountain View, but then he fell into an uncomfortable sleep.

The next morning George awoke feeling as if his body had turned to stone. His girlfriend Juanita was sleeping quietly on his shoulder, but when he tried to move his right arm he couldn't. It was stuck, wrapped around her body. Juanita pried herself free and helped George out of bed. Everything George did that day happened in slow motion—going to the bathroom, getting dressed, making breakfast. He had no desire to go out, but he had to show up in court or his parole would be revoked. If he failed to appear he would be back in prison before the day was out. Moving with glacial speed, George struggled out to his old Volkswagen and drove to the courthouse on Julian Street.

The courthouse guard noticed a strange figure shuffling past the metal detector and assumed that he was intoxicated. George never made it through the door. He was arrested on the spot for being under the influence, a parole violation. Within hours he was in the San Jose jail.

Each day in his cell, George's stiffness got worse. By the fourth day he could hardly move his arms. By the sixth, he could not talk. He could see people and hear them, he could feel the sensation if someone jostled him. But he couldn't turn his head or reply if someone called his name. He was terrified.

Finally a doctor was called. Struck by George's appearance, he immediately sent him to the emergency room at the county hospital, the Santa Clara Valley Medical Center. The emergency room doctors were skeptical. Prisoners will try anything to get out of their cells to a hospital, where the treatment and food (and the opportunities for escape) are much better. There was a good chance this patient was faking his bizarre condition. But there were other possibilities. Prison doctors sometimes give large doses of potent tranquilizers called neuroleptics to chemically restrain agitated or violent patients. On the chance that George had been overtranquilized, the emergency room staff took blood and urine samples and sent them off to the lab. Since the results would take time, they injected 25 milligrams of Benadryl—a drug known to reverse the effects of neuroleptics—in an effort to reverse the condition, and sent George back to prison.

But the Benadryl didn't work. Frustrated at seeing his patient no better, the jail doctor sent George back to the emergency room. This time the Valley physicians tried a stronger antidote, Cogentin, to overcome the effects of any tranquilizers George might have received in jail. It had no effect. They returned him to the jail frozen and mute. The next day both George and the tests came back. The tests were negative—there were no traces of tranquilizer in his blood or urine. Not having any idea what they were dealing with, but still suspecting that their patient was desperate to get out of jail, the emergency room physicians decided to push even harder to see if their immobile patient was faking his bizarre condition.

First they scraped the soles of his feet with the pointed end of a reflex hammer to jolt him out of his state. No response. They applied blunt pressure to the base of his fingernails (an excruciatingly painful maneuver that does not cause any tissue damage). Most people jump up howling with pain when this is done, but George remained motionless. Finally, in exasperation, they tried ammonium sulfate (smelling salts), which is so pungent it can raise a patient out of a dead faint. They broke a capsule and held it up to George's nostrils, but again—no response. Their mysterious patient remained absolutely frozen.

Inside, George was consumed with anger. He could hear everything they said. He could feel everything they did to him. He had screamed inside when they jabbed the soles of his feet, and felt like throwing up when they passed the ammonia under his nose. At one point he was so angry that he tried to hit one of the doctors. He willed his arm to move, and it did start, but it moved so slowly that nobody in the room noticed.

Satisfied that he wasn't a malingerer but having no idea what was wrong with him, the emergency room physicians sent George to the place where such unexplained disorders are sent—the psychiatry ward. The admitting psychiatrist made an initial diagnosis of catatonic schizophrenia, a rare condition in which an individual becomes rigid and mute as an emotional response to some devastating crisis. The diagnosis seemed reasonable to the resident who made it (who had never seen an actual case), but the senior psychiatrist, who had some experience with catatonic schizophrenia, decided the diagnosis just didn't feel right. A catatonic schizophrenic from jail? No, more likely there was a physical basis for the patient's condition. George needed to be seen by a specialist in physical disorders of the brain, a neurologist.

The neurologists had never seen a patient anything like George either. They discussed his case at length, standing over his frozen form. Trapped inside his body, George watched helplessly as the academic arguments flew back and forth. Maybe the heroin he had bought in Mountain View was to blame, but there was no way he could tell them.

Various doctors came and went. They prodded him, pricked him with pins, banged him with reflex hammers, and shone lights in his eyes. After a few days, two aids came into his room pushing a gurney: George was to be transferred to a special unit.

Watsonville, California, July 4, 1982

David Silvey was hungry, thirsty, and frightened. For several days he had lain motionless in his apartment, unable to move. Somewhere in the apartment was his brother Bill. He could hear him breathing but he couldn't move his head to see him. David was more scared than he had ever been in his life. He made a huge effort to lift himself up but nothing happened—his limbs just wouldn't move where he wanted them to. He tried to shout aloud, but no sound emerged from his dry lips.

If someone didn't come soon, they would be finished. Even if someone did come and ring the bell, the odds were they would leave when they got no answer. The telephone started ringing. Maybe it was his mother; maybe she would suspect that something was wrong.

David and Bill had had a good couple of days earlier in the week dealing a new shipment of heroin, and had decided to use the rest themselves. It was terrific stuff, unlike anything David had used before, although his veins burned when he fixed. They had gone on a run with their friend Big Mike, taking it for a solid five days, and it was great. But as the week wore on, David had felt himself slowing down. Big

Mike didn't seem to be much affected and left the apartment, but Bill and David had not felt like going out. In the end they stayed home watching TV. Now, on the Fourth of July, David was not even capable of changing the channel. As the phone stopped ringing, he drifted off into a strange sleep.

The next morning their mother arrived to find Bill and David lying there, totally paralyzed. She called an ambulance which took them to the Watsonville Community Hospital. The admitting physician could see no obvious reason why two apparently healthy young men in their twenties should become rigid and frozen, and he listed a diagnosis of catatonic schizophrenia. The problem, he concluded, was probably in their minds.

Department of Psychiatry, Stanford University Medical Center, California, July 5, 1982

Connie Sainz was in pain. Unable to move, she just lay in the bed surrounded by doctors and nurses. They asked her questions, she tried to answer them, but no words emerged from her lips.

Connie's nightmare had begun a month before, when she was resting in her apartment in San Jose. First Connie had felt tired, then stiff, then rigid. It was as if some evil spell were slowly encaging her in her body. By the time her sister Stella returned from work, Connie was lying completely immobile with her eyes wide open, unable to respond or talk. At first, Stella thought that it was the effects of heroin withdrawal. That summer, much to the consternation and shame of her family, Connie had started using heroin. After lots of family arguments, she had agreed to try and kick the habit. Perhaps this was the gruesome first effect of withdrawal.

After three days, Connie was worse. Her face lost all expression and now looked like a mask. Her body became twisted and contorted and underwent sudden jerks. By the tenth day, she looked more like a marble sculpture than a living woman. Stella had no idea what to do. In desperation she lifted Connie into the backseat of her car and drove north on Route 101. She had heard that Stanford had some of the best doctors in the world. That was where she would take Connie. Surely they would be able to cure her.

The emergency room doctors at Stanford had seen nothing like this and carried out a series of tests. Unable to discover a physical reason for Connie's condition, they focused on a detail in her medical history that Stella had provided. Apparently Connie's problems had begun after word reached her that her boyfriend Toby, a Salinas drug dealer, had become paralyzed. Perhaps this was a case of sympathetic paralysis—Connie had been so affected by the news of Toby's paralysis that she had undergone a hysterical conversion reaction, developing paralysis in sympathy. The problem was, as they put it, "all in her mind." This line of reasoning led them to admit Connie to the psychiatric ward, where she was subjected to a number of tests, including sodium amobarbitol (truth serum), to try to get to the bottom of her hysterical paralysis and what they saw as her "deeply rooted psychological problems."

To Connie, trapped inside her body listening to all this earnest theorizing, it was living a nightmare. She knew what she felt, but she couldn't communicate with the doctors. There was absolutely nothing wrong with her mind; the problem was that her body no longer did what her mind wanted. To make things worse, she was in terrible pain. She had been stuck in a frozen position so long that one of the nerves in her right leg had been crushed.

After two weeks of frustration, the psychiatrists finally gave up in their therapeutic attempts and sent her home with a diagnosis of "functional paralysis." Her mother was told that Connie would eventually get better; the problem was all in her mind.

Part One

■ ■ ■ ■

1

All in the Mind

Santa Clara Valley Medical Center Neurobehavior Unit, San Jose, California, July 16, 1982

The patient lay there frozen. Looking more like a mannequin than a person, he made no movement and uttered no sound. His face was expressionless and his eyes stared eerily ahead. Dr. Phil Ballard, the physician on duty, flipped through the patient's chart. He was a 42-year-old Hispanic man named George Carillo who had become catatonic two weeks earlier, shortly after being incarcerated in the San Jose jail. Since then he had been seen by a dozen or more doctors in different parts of the hospital, but it was clear that none of them had the slightest idea what was wrong.

The case notes told a bizarre story. Carillo had turned up for a court hearing intoxicated on PCP and was jailed on the spot. But in his cell, he became less and less able to function. For three days he had bounced back and forth between jail and the Valley emergency room before finally being admitted to the psychiatry unit. Within the department of psychiatry there was a specialized ward called the neurobehavior unit, designed to deal with unusual cases which did not seem to fit classic psychiatric classifications and where an underlying physical or "organic" cause was suspected. And so, on July 16, the mysterious case of George Carillo had been sent to them.

Ballard, who was doing a fellowship in behavioral neurology on the unit, put the chart down and looked at his patient. George Carillo hadn't moved the entire time that he had been reading the medical records. If his new patient was faking it, he sure was some actor. Ballard looked at his watch: 10:00 A.M. His boss Bill Langston would be swamped with patients in the general neurology clinic, and would not want to be dragged out. But George needed a diagnosis, and the sooner the better.

General Neurology Clinic, Valley Medical Center

That Friday morning the neurology clinic was especially busy. A dozen patients filled the small waiting room and another twenty lined the corridor outside. Bill Langston, the head of the department of neurology, was used to it. Already that day he had seen two stroke victims, one with frontal lobe syndrome, two Alzheimer's patients, a case of temporal lobe epilepsy, and three patients with migraine headaches. It wasn't glamorous or well-paid work, but it was important. For tens of thousands of poor people in Santa Clara County, some fifty miles south of San Francisco, Valley Medical Center was the only place they could go for medical treatment.

And it was interesting. Given the complexity of the brain, neurological cases are often difficult and challenging. While sophisticated imaging systems exist to image the brain, they are not sensitive enough to detect small lesions. With the exception of the nerve fiber layer in the eye, which can be seen through an ophthalmoscope, the nervous system cannot be observed directly. A dermatologist can see a rash, a cardiologist can listen to a heart, but a neurologist must localize any damage to the brain through a complex process of deduction and inference from the clinical signs and the patient's symptoms. For example, paralysis on the right side of the face and weakness of the left arm and leg means that the lesion—whether it is a stroke, tumor, or something else—must be located at the level of the facial nucleus on the right side of the brain stem. A lesion in this location can knock out the nerve fibers that innervate the muscles on the right side of the face (which are uncrossed), and at the same time damage motor tracts that run from the right side of the brain to the left side of the body, before they have crossed to the left at the junction of the lower end of the brain stem and spinal cord.

The symptoms vary greatly from patient to patient. One of Langston's patients had been walking down the street one day when he started to hear the voice of Walter Cronkite introducing the *CBS Evening News* and wondered if he was losing his mind. Neurological evaluation revealed that his auditory hallucinations were the first sign of a brain tumor in the speech area on the left side of his brain.

For all this, Langston had a sense that his career was going nowhere fast. At age 39, married for the third time, with a baby on the way, it looked increasingly unlikely that he would achieve major recognition as a physician, much less as a researcher. For some years he had held a teaching position at Stanford, but the chances of achieving tenure were virtually nonexistent. To get tenure it wasn't enough to be a good clinician and teach, you had to do significant research and publish papers. And Langston just didn't seem to have time to do much research.

Since his early teens, Langston had been fascinated by human behavior and its underlying physical causes. When people felt hunger, anger, pleasure, or pain, what was going on in the brain? Why did one person become an alcoholic or another an obsessive-compulsive and yet another a manic-depressive? When psychiatrists spoke about such things it was often in terms of Freudian psychodynamic theory—which Langston had always found fascinating, but also somewhat circular.

In striking contrast was the way scientists studied behavior in animals. As a boy, Langston had watched programs on public television showing experiments in which rats would push a lever that electrically stimulated a certain area of their brain thought to induce a feeling of pleasure. The rats would push this lever rather than one that provided food, to the point of starvation. Langston had been intrigued by such experiments. These scientists were attempting to dissect the neurological basis of behavior, to discover the physical basis of pleasure and hunger. Perhaps such experiments on animals would eventually shed light on why people behave the way they do.

Langston had gone to medical school and, after toying with the idea of going into psychiatry, specialized in neurology, thinking that clinical neurology was a sound background for doing research into human behavior and the brain. But somewhere along the line, he had become virtually a full-time clinician. Now his days were so filled with the practice of medicine that he had little time to even think of research, let alone carry it out. Despite his expertise in diagnosing and treating patients, he felt unfulfilled. From time to time he wondered whether he should have studied basic science rather than medicine and set out to discover something new. He had always wanted to make a difference, but now it just didn't seem to be in the cards.

Langston was just about to start on a new patient when a call came through from Phil Ballard. "You must come and see this patient who was just admitted to the neurobehavior unit, Bill, you're just not going to believe it!" Langston listened impatiently as Phil outlined the case and how the psychiatrists had decided that the case belonged on Langston's neurobehavior unit. It sounded interesting, but not exactly urgent. Phil Ballard was insistent. "Absolutely. I'm telling you you must come right now. The psychiatrists don't know what to make of him."

Langston started to walk to the neurobehavior unit. In a sense, this unit was Langston's research. Two years before, he had set up this small unit in the psychiatry department to diagnose and treat unusual cases. It was well known that from time to time cases turned up in psychiatry that were not truly psychiatric patients. These patients' disorders were caused by specific and sometimes irreversible damage to their brains, and therefore they needed to be seen by a neurologist. While these patients needed to be properly diagnosed and treated, they also offered a research opportunity. Such cases might give Langston a way of linking behavior to physical changes in the brain.

To date, the unit had admitted some interesting cases. There had been a young college student who had begun preaching in class and then aggressively approaching female classmates. He proved to have Wilson's disease, a rare disorder of copper metabolism. There was a patient who had started hallucinating wildly and drinking water from a toilet bowl, who turned out to have a herpes encephalitis, a viral infection that affects the temporal lobe of the brain. In both instances, abnormal behavior had resulted from organic changes (lesions) in their brains. Langston liked to call such cases experiments "of nature" rather than of science. Neurologists could not study humans the way neuroscientists studied rats in the laboratory, by stimulating or removing parts of the brain and observing the change in behavior. They had to wait for cases to arise naturally. But when they did arise, they offered scientists a small window into how the brain operates and controls behavior.

Wilson's disease, viral encephalitis, and today a frozen man. Langston felt sure that important insights would be gleaned from these and other experiments of nature transferred to his unit, but they hadn't yielded any breakthroughs yet.

Yet the minute Langston walked into the room and saw the mystery patient, he understood why Phil had called him out of the clinic. George Carillo lay propped up in bed staring straight ahead. His mouth was slightly open and drooled continuously. His arms were bent at his side with the elbows outward, as if frozen midway through a motion. He made no sound.

Langston also immediately understood why the psychiatrists had not been sure what to make of George. The zombie-like appearance did look like catatonia. Langston took George's arms and very gently pulled them toward him. They were very stiff. Langston held the arms straight out in front of George's body and let them go. They just stayed there fixed in space where he had left them. After thirty seconds they began to fall slowly and over the next three or four minutes returned to George's side. Next Langston lifted one of George's arms above his head and let go. Again it stayed there for thirty seconds or so before slowly falling under gravity. Langston had seen this before. It was called "waxy flexibility" and it was a legendary sign of catatonia. He saw why some of the psychiatrists had opted for this diagnosis.

He turned his attention to George's face. It was totally expressionless—more like a mask than a face. The eyes stared eerily ahead without blinking. Langston gently tapped the forehead between the eyebrows. This will cause anyone to blink, even someone in a catatonic state—it is an involuntary reaction to the stimulus of tapping. However, with repeated tapping most people stop blinking. But George didn't. The failure to stop blinking is known as Myerson's sign and is typically seen with damage to an area of the brain known as the basal ganglia.

On several occasions, George's eyes slammed shut and would not reopen. "Try and open your eyes, please, Mr. Carillo," Langston asked whenever this happened. He had no idea whether George could understand him or not, but thought it better to presume competence on the part of his patient. For a while nothing happened. Then, after some thirty seconds, George slowly opened his eyes. This condition is called "eyelid apraxia," and it is very rare—so rare, in fact, that it is not definitively associated with any known neurological disease. Langston had no idea what it meant in George's case.

While Langston had been examining George, the room had filled up with physicians—mostly psychiatrists, but also several neurology residents. George had created quite a stir in the psychiatry department, and there was an ongoing debate as to whether his condition was psychological or physiological in origin. The psychiatrists had become convinced that George's condition was organic or neurologic in nature, whereas the neurology house staff had concluded with equal vehemence that his condition was entirely psychiatric. It was now up to Langston to make the final pronouncement.

The waxy flexibility indicated catatonia, a psychiatric condition. But the Myerson's sign and the eyelid apraxia pointed toward a neurological cause. Langston continued his examination of George's face. The upper part of the face was unusually oily, a condition known as seborrhea. Seborrhea, like Myerson's sign, is seen with certain diseases of the basal ganglia.

Langston gently took George's arm, holding the wrist in one hand and supporting the elbow in the other. Slowly he tried to bend the arm. It was very stiff. Langston had to apply considerable pressure to make the joint move, and when it did it moved in fits and starts, like a ratchet wheel. Langston relaxed. He was now certain that George had a neurological disorder. He had seen several cases of catatonic schizophrenia on the neurobehavior unit and they didn't show any "cogwheel rigidity."

It appeared as if George had incurred damage to the basal ganglia area of his brain. But how? The blood and urine tests had ruled out tranquilizers as the cause, and there was no evidence of infection.

The most important element of any medical diagnosis is the history—the tale of the patient's illness, life, work, and previous illnesses. But with George they were stuck: not only could he not speak, but it was highly likely that his memory and reasoning abilities were also impaired, that his mind was as frozen as his body. They would have to proceed cautiously and hope for a break.

George was now officially admitted to the neurobehavior unit and put under close observation. Several times a day, nurses repositioned him in bed so that he wouldn't develop bedsores. They washed him and fed him. Despite his external appearance, his internal organs appeared to function normally. If food and water was administered in small teaspoons into his mouth, he was able to swallow it. He also seemed to have control over his bladder and bowels.

For seven days George lay there without speaking or moving. Then one morning Phil Ballard noticed him moving his fingers ever so slightly. The movements were slow and looked as if they could be voluntary. On a long shot, Phil gently wrapped the fingers of George's right hand around a pencil and slipped a yellow notepad under it.

"Write your name," Phil said. Unexpectedly and very slowly, the pencil started to move. After a minute it was clear that their patient was trying to write something, perhaps the first letters of a name: *"G . . . e . . . o . . .* Five minutes later he had completed a name, *George Carillo.* After half an hour, there were three more sentences on the page of paper: *I'm not sure what is happening to me. I only know I can't function normally. I can't move right. I know what I want to do. It just won't come out right.*

Langston and Ballard were astonished. Trapped inside this frozen body was a normal mind. Langston thought of the awful process George had been through: being prodded, pricked and scraped, having ammonia stuck under his nose, feeling pain but being unable to scream, hearing questions but being unable to respond. Never had there been a better illustration of the warning all doctors are taught from the beginning of medical school: No matter how bad a patient's mental status seems, whether it be deep coma or utter confusion, never carry on clinical discussions at the bedside on the assumption that the patient can't hear you. Langston could only wonder what discussions George must have listened to during those fateful emergency room trips. But at least now they had a window to his mind, a way of getting a case history, and so by question and slow answer they painstakingly gathered evidence.

George told them that about two weeks earlier he had started feeling stiff and that his girlfriend, Juanita Lopez, was having the same problem. Probing for clues, Langston asked him to write down any medicines he was on. In reply, he didn't write down the name of any prescribed medicine. Rather, he wrote down the word "heroin." Langston looked at Ballard and smiled. "No, George, I don't mean street drugs, I mean medications that doctors prescribe." George showed no reaction.

A little research revealed that George's girlfriend, Juanita Lopez, 30 years old, had been living with George before he was incarcerated and was now staying with her sister. But when Langston called Juanita's family, they seemed reluctant to talk to him. They did, however, confirm that Juanita was sick.

After days of cajoling and pleading, Langston managed to convince Juanita's family to bring her in to Valley Medical Center to be examined. Langston was shocked at what he saw. She sat completely motionless, like a wax doll. Her face was acned and, like George's, expressionless. Her eyes hardly blinked. She drooled continuously. For two weeks now her family had been taking care of her in this condition. In the morning they would get her up, bathe, dress, and feed her, then put her in a chair for the duration of the day. In the evenings this process was repeated in reverse before they put her back to bed. In short, she required the type of total body care that is usually found only in hospitals. The family was desperate for help.

They confirmed that shortly after Juanita returned from George's place on July 1, she had frozen up completely. They had taken her to a local hospital but, thinking she was high on PCP, the hospital had turned her away. Her family, worried that the police might then arrest her, had decided to keep the matter to themselves. Langston could sense the relief they felt now that they had brought her in. He could also sense their expectations that he would make her better. But Langston knew he could do little until they discovered the cause of the strange symptoms.

Juanita's clinical history matched George's. Whatever had gone wrong, the symptoms for both of them began on

July 1. Like detectives at the scene of a crime, Langston and Ballard searched for clues. The obvious place to start looking was George's apartment. The cause of the mysterious paralysis might be something in their common environment—something that they both ate; a furnace leak of carbon monoxide or some other chemical. At this stage, Langston didn't want to rule out anything.

There was a very well-known neurological disease which seemed to fit their physical signs and symptoms in every detail: Parkinson's disease. Between 500,000 and 1 million Americans have this disease, including one of every hundred people over the age of 60. Parkinson's sufferers freeze up and have difficulty moving. Their limbs exhibit cogwheel rigidity and their speech becomes soft and slurred. Advanced cases have expressionless "reptilian" faces (as they are called) and oiliness of the skin. But Langston had two problems with this diagnosis: Parkinson's disease doesn't strike people overnight, and it hardly ever affects the young. It usually occurs after the age of 50, and comes on very gradually, so gradually that most patients have symptoms for one to two years before they go to a doctor. But the Valley Medical Center cases were young—Juanita was barely 30—and their symptoms had developed in days rather than years. Langston no longer thought about his career rut. He had a first-class medical mystery on his hands.

The weekend after Juanita came in, Phil Ballard decided to take off for a weekend break. He drove thirty miles south on Highway 17 to Santa Cruz, where he had been invited to a party. The host was Dr. Jim Tetrud, a Watsonville neurologist in private practice. As the evening wore on, the two neurologists began talking about the subject that drove their professional lives, the study of the human brain. Phil was on the verge of describing the mysterious cases he had seen at Valley Medical Center, but before he could, Jim Tetrud told Phil an equally astonishing tale of two patients he had seen that week in the Watsonville Community Hospital.

Two brothers, David and Bill Silvey, had been found lying

frozen in their apartment, unable to move or talk. On discovering them, their mother had called an ambulance which took them to the Watsonville Community Hospital. The admitting physician, seeing no obvious medical reason why two apparently healthy young men in their twenties should be frozen, listed a diagnosis of catatonic schizophrenia.

The family's private physician, Dr. Sean Murphy, was bothered by the diagnosis. Two brothers developing a rare condition like catatonia at the same time? He decided to ask Dr. Tetrud to take a look at David and Bill. Both brothers had a long history of drug abuse and apparently had been taking drugs continuously for more than a week before they froze up, so initially Tetrud suspected that the symptoms were caused by intoxication. Tetrud gave them Benadryl and later Cogentin to unfreeze them, but the medicines were totally ineffective.

"It's incredible," Jim Tetrud said to Ballard, "but it's almost as if these young men have advanced Parkinson's disease."

Phil Ballard couldn't believe what he was hearing. "Did you say they were drug addicts? What was their favorite drug of abuse?"

"Oh, heroin, I think," said Tetrud.

As Jim Tetrud sketched out the symptoms of David and Bill Silvey, it was as if he were describing George and Juanita. Yet these were two different people from a completely different location. Ballard got on the phone to Langston and repeated what Tetrud had said. When he had finished, Langston asked him, "Do the Silveys have any relationship with George and Juanita?"

"No, Bill, as far as we know they have never met."

Langston paused. "We'll have to check on Monday that there's no connection, but assuming that's true, assuming they didn't know each other and had never visited each other's apartment, then there's only one connection between the four cases, isn't there?"

"Yes," said Phil, "they are all heroin addicts."

After hearing about the Silvey brothers, Langston found

it impossible to relax. Four young people's lives had been destroyed in a tragedy that seemed to hinge on drugs. His mind filled with a far-fetched theory. Perhaps something was not right about this particular batch of heroin, and when injected into the bloodstream it had passed into the brain, damaging it so as to produce the symptoms of advanced Parkinson's disease. As far as Langston knew, no such substance had ever been encountered before. There were drugs that caused transient parkinsonism—neuroleptics like chlorpromazine and haloperidol—but George's emergency room doctors had tested his blood and urine for these and found nothing.

As Langston wrestled with the medical mystery, a chilling thought crossed his mind: If the heroin was to blame, northern California might have a major public-health disaster on its hands. They had seen cases in San Jose and cases fifty miles away in Watsonville. What if this batch of heroin was being sold all over northern California? Two things needed to be done immediately. First, Langston needed samples of the heroin that the four addicts had used to send for chemical analysis. Second, he had a duty to warn the public that there was a poison on the streets, and that anyone using heroin was in grave danger. Otherwise there might be an epidemic of frozen addicts.

Langston had never had any dealings with the press, and consequently had no idea how to get a message out to the public. The medical director of Valley Medical Center referred him to the hospital's public relations firm, PRx, which in turn put out a brief release to local newspaper and television reporters. PRx, which generally had a hard time generating any interest in Valley Medical Center among the local media, was astonished. There were so many inquiries for further details that PRx decided to hold a press conference the next day and get Langston to speak.

By ten o'clock the next morning, interest had spread to the television stations from the San Francisco Bay area. The administrative conference room of the hospital, which had been hastily converted to a press room, was packed with cameras and reporters. As he had no idea how to talk to the press, Langston spoke as if they were medical students and he was conducting morning rounds. He made a brief statement, answered a few questions, and left, glad he had gotten the message out.

Sound bites from the press conference made the nightly news on four of the northern California stations, which meant that, at that time of year, approximately twenty million people heard about the story. Even though drug addicts weren't known for their patronage of the nightly news, it was likely that a sizable fraction either saw or were told about the broadcast.

In San Jose, a physical therapist named Jan Bartell happened to be watching. As she looked at the videos of George Carillo, she became increasingly agitated. She wrote down Langston's name and resolved to call him the next morning.

For several weeks she had been making home visits to a young woman named Connie Sainz with a very unusual "mental" condition. About six weeks earlier Connie had frozen up. Over a couple of days Connie had become a total invalid and now she couldn't walk, talk, or do anything for herself. Connie had spent two weeks on a Stanford Medical Center psychiatric ward undergoing tests, but had been discharged when doctors decided that her "paralysis" was hysterical in nature. In such patients the paralysis usually went away without intervention, and home-based care seemed the appropriate solution.

But Connie was not doing well, and after seeing the television pictures of George Carillo, Jan began to wonder whether Stanford's doctors might have made a terrible mistake. The patient on TV looked uncannily like Connie. He had the same reptilian stare, the same flat facial expression, the same paralysis. The next day she called Langston at Valley Medical Center and told him everything. Langston was very excited and asked Jan to bring Connie and her family in immediately. The press conference had already brought totally unexpected results.

Even though Langston had seen four cases of this drug-induced state, he was still shocked when he saw Connie. Just 21 years old, this young woman was a tragic sight. She had bedsores from lying in the same position for six weeks, and a crushed nerve in one of her legs. She couldn't do anything for herself. She found eating difficult and had to be hand fed. She couldn't move her arm to pick up a cup of water. She couldn't stand up by herself. When Langston put his head close to her chest he could just hear a whimper, but otherwise she could not talk. Sometimes Connie's eyes would close in reaction to a noise or a doctor's hand that was examining her. If she was left to herself, it took her forty seconds to reopen them. Her symptoms closely matched those of George, Juanita, and David and Bill.

Yet Connie's story, as told to him by her family, touched Langston in a way that the others' had not. She came from a poor but educated Hispanic family in Greenfield, a small town in California's central valley surrounded by miles of hot, dusty fields. Until 1982, Connie had had a fairly normal life. She had finished high school and gone to work. Though not married, she had one son. And she had managed to avoid drugs. But then she had met Toby.

Toby Govea, soft-spoken with gentle eyes and a sweet smile, was a dealer who had repeatedly been in trouble with the law. Drugs were an everyday activity in Toby's family—he routinely shot up with his father, brothers, and sisters. Whereas a middle-class family might discuss the day's events while the parents sip martinis, Toby's family would chat as they shared heroin, PCP, or cocaine. Toby not only used heroin, he sold it on a large scale, and to support his habit he robbed stores.

Toby later claimed that he had urged Connie not to use the drugs he was selling, but if he did she ignored his advice. One day in June 1982, perhaps saddened because Toby had been taken off to the Salinas jail, she started fixing with some new heroin that Toby had been selling. She used it every day for a week. Then she heard from a friend that Toby had fallen ill with a strange freezing disease. After a few days, she became listless and apathetic, moving less and less. Then her body became twisted and contorted, her limbs moving into strange positions in fits and starts, at times jerking violently. After the movements stopped, Connie began to stiffen up and soon couldn't move or talk. Her face lost all expression and became like a mask. By the tenth day she looked more like a marble sculpture than a young woman. Connie's sister Stella had been so worried that she had taken Connie to Stanford Medical Center. But whatever was wrong with her, Stanford hadn't discovered it.

As Langston subsequently discovered, Toby was transferred to Stanford as well. Like Connie, he was frozen, but also had a pronounced parkinsonian tremor. Stanford's neurologists didn't know what to make of him, but in the course of his examination it was discovered that he had been briefly treated in prison with phenothiazene, a class of neuroleptics that can induce temporary parkinsonism. Since they had no other ideas about what was wrong with him, they concluded that neuroleptics were the cause of his bizarre condition.

As a Stanford faculty member, Langston was deeply chagrinned when he heard this story. One of the neurologists treating Toby was a former student of his. But that wasn't the end of it. The neurologists in this case went so far as to make a videotape of Toby, which was then placed in the Fleischman Library (an innovative video teaching library for medical students at Stanford) as a classic example of phenothiazene toxicity. To this day, if a medical student goes to this audiovisual medical teaching library to learn about phenothiazene-induced parkinsonism, he or she will see Toby. It is somewhat painful to read the message at the end of the tape: "This is a case of phenothiazene toxicity. Since the medication has been discontinued, he will eventually return to normal." But Toby did not have phenothiazene toxicity, nor would he ever be normal again.

As July turned to August, Langston knew of six cases: George, Juanita, David, Bill, Connie, and Toby. Three were now being treated at VMC, two in Watsonville, and one at Stanford, but he had assumed de facto responsibility for all of them. The mystery of their condition was consuming an enormous amount of his time and energy. But it was to solve this kind of fascinating mystery that he had gone into neurology in the first place.

Because the key to the mystery centered on the heroin, Langston and his colleagues now focused on obtaining samples for analysis. By coincidence, the day after the Silveys were taken to the hospital, the Watsonville police had raided their apartment. They didn't find the Silveys, but they did find several bindles of heroin. When Langston contacted them, they were very cooperative and sent a squad car up to San Jose with sirens blazing to deliver the sample to Valley Medical Center. A search of George's apartment had turned up a gram of heroin in his refrigerator. For completeness, Langston would need a sample of the batch that Connie and Toby had taken.

Langston sent samples of the heroin to various toxicology labs in northern California for analysis, and tried to get on with his life while he waited for the results. Every day he called the labs to get a progress report. They told him that whatever it was, it wasn't heroin; it was a synthetic concoction bearing no relation to the opium poppy. It didn't make a lot of sense to Langston that addicts would call a synthetic substance heroin if it wasn't heroin. So what exactly was this synthetic substance? The labs were not sure.

Ian Irwin of Stanford's Drug Assay Laboratory successfully isolated a pure extract from the synthetic heroin and fed it into a gas chromatograph/mass spectrometer, an instrument that uses a stream of electrons to shatter molecules into little pieces. The way the molecules break apart and the pattern formed by the fragments form a molecular fingerprint. Chemical compounds have fingerprints every bit as unique as the ones people have. To a chemist, a chemical like codeine, cocaine, or salt is completely recognizable by its chemical fingerprint. The fingerprints of some forty thousand known chemical compounds were stored in a database in Washington, D.C. Irwin's strategy was to first fingerprint the new heroin and then screen it against the forty thousand compounds to look for a match. But after considerable time and effort, no match could be found. He could not identify what was in the synthetic heroin.

Meanwhile, the condition of the six addicts was deteriorating. They were invalids, unable to wash, dress, or feed themselves. George and Juanita had become so ill that Langston began to fear for their lives. His biggest fear was that one of them might develop a pulmonary embolism—a blood clot that forms in the leg and passes into the lungs, where it blocks an artery and causes sudden death. Any patient who is bedridden and completely inactive is at high risk of this. Langston was also worried about infection. Some of the patients, notably Connie, had developed serious bedsores. If these open sores were not kept clean they might become infected; if the patient became septic (blood-borne infection) it could easily be fatal.

To make matters even worse, these patients had all lost a lot of weight and were very debilitated and vulnerable to infection. If one of them developed pneumonia, they might not be strong enough to fight it. Remarkable though it was, these six young people appeared to be dying of Parkinson's disease.

Parkinson's is one of the two great neurodegenerative diseases of aging (the other, Alzheimer's, entails progressive loss of memory and other mental faculties). Past victims of Parkinson's are thought to include Mao Ze-dong, Francisco Franco, and Adolf Hitler. It has been known for more than fifty years that the symptoms of Parkinson's result primarily from the death of nerve cells in a small area at the base of the brain called the substantia nigra, which means "black substance." These cells produce a brain chemical called dopamine—a so-called neurotransmitter, which nerve cells use to communicate with each other—that is essential for normal control of movement. Without dopamine, the *thought* of lifting an arm can't be transferred into the *act* of lifting an arm. In Parkinson's disease, nerve cells (or neurons) in the substantia nigra die. As the number of neurons diminishes, so does the production of dopamine, and the motor system begins to shut down.

As parkinsonian patients become more and more disabled, they become highly susceptible to what physicians call "intercurrent illnesses," such as pneumonia and sepsis. After a period of being bedridden, the advanced Parkinson's disease patient can succumb to an infection or a pulmonary embolism. Langston now feared the same fate for his patients.

In the late 1960s neurologists began treating their Parkinson's patients with a new drug called L-dopa. The cells in the brain use this chemical to make dopamine. By being given the chemical precursor L-dopa, the remaining cells in the brain can be induced to make more dopamine, partially reversing the deficiency and giving patients back voluntary control of their muscles. As there was nothing to lose, on July 28, Langston administered the drug to George and Juanita and waited.

The effect was miraculous. Within hours, George and Juanita unfroze and came back to life. Within days they could walk normally, move normally, speak normally, and feed and wash themselves. As a medical student Langston had read Oliver Sacks's moving book *Awakenings,* in which L-dopa awakens a group of patients afflicted with a rare form of parkinsonism known as post-encephalitic parkinsonism. What Langston was seeing was scarcely less remarkable. The frozen addicts could now talk about the private hell they had all endured for weeks—the psychobabble, the mistaken diagnoses, the frustration, the fear, the pain, and the sadness. The four other patients, including Connie, proved equally responsive to L-dopa.

George told of how he had endured being prodded and scraped. Connie wept as she recounted the tale of her misdiagnosis at Stanford and how she had to listen to all the earnest theorizing about her mind, and how, despite the pain from the crushed nerve in her leg, she had been unable to cry out.

Here were six patients whose symptoms modeled, down to the last detail, one of the major unsolved degenerative diseases of the aging brain. What did it mean? In the weeks ahead it would become clear that what had started as a drug tragedy was to open a new chapter of medical research which would offer hope to Parkinson's disease sufferers throughout the world.

2

Drugs by Design

Santa Clara County, California, July 1982

Dave Weidler knew that he was beginning to smell. He hadn't changed his clothes for a week. He hadn't shaved or washed his hair in a month. He was hanging around with the wrong kind of people—drug dealers, pimps, petty thieves. But then that was his job. He was a cop.

For the past year, Deputy Dave Weidler had been working undercover as part of the Santa Clara County Police Department's continuing war on drugs. It was a war that was going badly for the county. Not only did they have to deal with traditional drugs of abuse like heroin, cocaine, marijuana, and PCP, now they had to cope with a new problem, "designer drugs."

It seemed to Dave that no matter how many laws were put on the statute books, the bad guys had an answer. When commercial production of the hallucinogenic drug lysergic acid (LSD) had been shut down in the mid-1960s, illicit chemists in the California area began synthesizing and selling their own LSD. Later, when Congress passed stringent laws to control the production of amphetamines, maverick chemists started making their own "speed." When pharmaceutical manufacturer Parke-Davis stopped making phencyclidine, "speed labs" began selling it under its more usual name—PCP, or Angel Dust.

Around 1980, one of these back-street chemists hit on a brilliant concept. Instead of trying to make existing illegal drugs like LSD and PCP, why not make entirely new derivatives of those drugs? Why not "design" new drugs that nobody had ever seen before? Such drugs would be hard to detect, difficult to analyze, and, to cap it all, completely legal. A designed drug could look like heroin, taste like heroin, and act like heroin, but as its chemical formula was new, it was beyond the law. Not being on any list of controlled substances (which are illegal even to possess), such a drug could be made, sold, and used with impunity.

Law enforcers like Dave Weidler had been thrown into disarray. How could they *control* drugs of abuse when they didn't know what to control? Catch someone making PCP and, with luck, they would go to jail. But there was no point arresting someone for making a new designer version of PCP, because it wasn't illegal. Before the police could act, this new drug had to be fully analyzed and added to a list of banned ("scheduled") substances—a process that could take years. Meanwhile, the designer chemists had plenty of time to come up with yet another completely legal chemical variant of PCP. Until Washington figured out a way of fighting designer drugs, agents like Dave Weidler had to improvise.

His undercover assignment had gone well. He had obtained a packing list of chemicals due to be delivered to a house in Morgan Hill: 50 gallons of ether, 40 gallons of acetone, 100 pounds of lithium metal, 88 pounds of bromobenzene, 220 pounds of propionic anhydride, and 44 pounds of 1-methyl-4-piperidone. What would a resident of Morgan

Hill want with such large quantities of chemicals? As Weidler wasn't qualified to decipher the list, he turned it over to Jim Norris, head of the Santa Clara County Crime Lab.

From the outside, the county crime lab is not much to look at: a featureless, single-story modern rectangular block. Inside it teems with activity, with forensic chemists, toxicologists, weapons experts, and dozens of others devoting their scientific skills to putting criminals behind bars. Jim Norris read carefully through the packing list. The chemicals were quite unusual and expensive, and the quantities involved were very large. At a rough guess, Norris figured they were worth ten thousand dollars.

Norris was sure that the list was the basis of a designer-drug operation. The most likely drug of manufacture was an analog of fentanyl, an extremely powerful anesthetic developed for surgery by Janssen, the Belgian pharmaceutical company. Being both potent and short-acting, fentanyl was ideal for major abdominal surgery. Surgeons loved it. So did designer drug makers.

In fentanyl, illicit drug manufacturers saw a drug that not only gave a heroin-like high but also was cheap and simple to make. In contrast to the elaborate process needed to process natural heroin from the opium poppy, fentanyl can be synthesized from a few ordinary industrial chemicals. But there was much more at stake. The fentanyl molecule is so versatile that by making tiny changes, a chemist can not only create a totally new, and totally legal, version of fentanyl (called an analog), but also increase its narcotic effects a thousandfold. Other small changes to the molecule can affect the duration of action. The fentanyl that surgeons use only lasts thirty minutes, not long enough for most addicts. But a small change in the drug's molecular structure—adding a methyl group (a collection of a few atoms)—makes the drug last four or five hours, like real heroin. Other changes can cause the drug's effects to last as little as five minutes or as long as five days.

Norris had been amazed at how these new young entrepreneurs could tailor their product for the street market, varying potency and duration. As a law officer, he wanted to put these people behind bars. As a scientist, he couldn't help but feel a sneaking admiration for them. At least some of the people making fentanyl were world-class chemists, many of whom probably had advanced pharmaceutical training. In the opinion of his colleague Garry Henderson, an analytic chemist at UC Davis who coined the phrase "designer drug," the quality control of some of the designer drugs appearing on the street was as good as that found in any pharmaceutical company. But what struck Norris as really extraordinary was the sophistication of their approach. It was clear that some of these chemists were highly familiar with the latest scientific research literature. In a few cases, there was evidence that the designer chemists had actually gone beyond the literature to invent entirely new ways of manufacturing families of chemicals, like the fentanyl series.

As a business, it was brilliant. Because some fentanyl analogs were so potent, only very tiny amounts had to be synthesized. The chemists would then mix the drug with lactose (milk sugar) or some other filler. The final powder that appeared on the street that looked and acted like heroin was in fact 99.99 percent filler and only 0.01 percent fentanyl compound. In this way, an investment of ten thousand dollars yielded about a kilogram of synthetic heroin worth a million dollars on the street. It was later estimated that a household closet full of one potent fentanyl analog would be enough to supply half of America's annual heroin consumption.

But fortunately for the police, the designer chemists were not beyond making mistakes. In one spectacular example, on January 12, 1981, there was an explosion in the garage of a home on Canyon View Drive in the peaceful and picturesque town of Saratoga, California, which is nestled in the foothills of the Santa Cruz Mountains in the western part of Santa Clara County. Firefighters noticed narcotics and called Dave Weidler in to help. Working together, they carefully searched the premises. It appeared that the garage had contained a fully operational laboratory with huge stores of chemicals. They found lab notebooks with meticulous laboratory records dating back to December 1980, indicating the synthesis of certain fentanyl analogs. The various articles and books, many of which were partially burned and had come from the Lane Library at Stanford Medical School, showed that extensive research had been done, even going so far as obtaining the patents for the manufacture of fentanyl. The most potent compounds were marked with yellow highlighter. Other articles were on the synthesis of different drugs, including analogs of PCP and Demerol.

The owner of the house was very cooperative and admitted that he had accidentally started the fire. But he claimed that the garage was leased by a man named Vincent Mason, a lawyer. Weidler interrogated Mason and got nowhere. Because no laws had been broken, there was nothing more they could do, except to close down the garage laboratory as a safety hazard.

Mason, now living in Morgan Hill, was the recipient of the long list of chemicals that Weidler had just given Norris.

A source had told Weidler that Mason had purchased the chemicals to make designer narcotics, but, fearing retribution, this tipster would not testify against him. So Weidler, Jim Norris, and district attorney Doug Southard pondered their options. What they needed was a search warrant to see what Mason was up to. But how could they convince a judge to grant one? The chemicals on the packing list were perfectly legal. The substances he was trying to make were in all probability legal. As far as the law was concerned, he was a law-abiding citizen. But maybe there was a way. Storing such large quantities of flammable chemicals in a private residence might well be illegal under the fire code of the state. So District Attorney Southard suggested they enlist the help of the fire department.

The fire department was pleased to cooperate and staged

a surprise public safety inspection. A cleaned-up Dave Weidler went along, posing as one of their team. When they arrived at the house they were met by a man in his fifties. He was calm and polite and cooperated fully with the fire department. When asked what he intended to use the chemicals for, Vincent Mason replied without batting an eye, "I'm experimenting to develop new sno-cone flavors and moisturizing creams." The chemicals, most of which were poisonous and highly flammable, were obviously unsuitable for this purpose, but there was nothing Dave Weidler or the fire inspectors could do.

Unfortunately for Weidler, their timing was off; no chemistry was going on the day of the inspection. The glassware was set up, but it was clean. Weidler had brought along a special field presumptive test kit to screen for illegal substances like PCP and heroin, and while Mason wasn't looking he tested some powder traces on the bench. If just one sample tested positive, Weidler knew he would be able to get a full search warrant. But everything was negative. Feeling very angry and frustrated, Weidler covertly removed a sample of powder for later analysis by the crime lab. The fire inspectors gave the owner eight hours to remove all the chemicals from the premises. When they returned for a repeat inspection a few days later, both Mason and the chemicals had vanished.

Back at the crime lab, Jim Norris passed the powder to toxicologist Halle Weingarten for analysis. She quickly decided it was not fentanyl and contained no PCP. After some analysis, Weingarten concluded that it was closer in structure to the painkiller meperidine, better known as Demerol.

For Norris and Weidler, it seemed like the trail had gone cold. Despite all his undercover work, Weidler didn't know if they had prevented a new designer-drug operation or arrived as one was shutting down. Norris too was disappointed, but he had plenty of other unsolved cases to occupy his attention and put this one out of this thoughts.

The technicians and scientists at the crime lab took their coffee breaks in a large room at one end of the long building. There they chatted about their work, their lives, and the stories in the news. A few days after the fire inspection, there was a lot of talk around the lab about a bizarre local drug tragedy. Just miles away, at Valley Medical Center, a group of drug addicts had turned up with a strange Parkinson's-like condition and the neurologist in charge, Dr. Langston, had been on television warning drug addicts there was some very bad "heroin" on the street.

Norris had seen the videotapes of the frozen addicts on television and wondered vaguely whether there could be any connection with his designer-drug investigation. It was a long shot, but perhaps some of Mason's chemistry had gone badly wrong. *If* Mason was behind it, and *if* they could trace the drug that George Carillo and the others used back to him, then maybe they could catch him after all. Norris made a few phone calls and discovered that the Watsonville police had found samples of the synthetic heroin and that all of those samples had been sent to Dr. Langston.

Out on the streets, San Jose's addicts had their own theories about what had happened. The most macabre theory, which many addicts believed, concerned a recent break-in at a mortuary. Earlier that summer, thieves had broken into a mortuary in Monterey and stolen some embalming fluid. The thieves, so the story went, had "cut" some heroin with the highly toxic embalming fluid, and it was this that had caused George and the others to freeze up. Indeed, street addicts had their own name for the strange frozen condition: "the walking death."

At Valley Medical Center, Langston was still waiting to hear some definitive news from one of the laboratories. In between seeing patients and teaching, he was tracking a sample of the heroin Connie Sainz had taken. According to Connie, she had shared her heroin with a friend who had taken the remaining bindles home in a suitcase. Langston painstakingly tracked down the friend's address and phone number. He wasn't able to contact the friend, but did talk with her mother, who told him that after hearing what happened to Connie, her daughter had become hysterical and had gone to the emergency room at Salinas. Her daughter was together enough, however, to take a sample of the poisonous drug with her, assuming that if the doctors knew what it was they might be able to save her from Connie's dreaded fate.

Langston spent an hour trying to locate the physician who had been on duty, only to learn that he was on a month-long vacation. Finally he reached the supervising nurse who had been on duty that day, and she remembered the incident clearly. She told Langston that rather than analyzing the sample, the doctor had given it to the police.

It took Langston several days to track down the Salinas police officer who had received the "heroin." After checking his records, the officer called Langston back to tell him what had happened. It seemed that on receiving the drug, the officer had called up Stanford, where they were treating Connie Sainz, to see if they wanted it for analysis. He did this thinking that it might somehow help Connie.

"But that's great," said Langston.

"I'm afraid not," the officer replied. "The doctor I reached explained that Connie had a psychiatric illness and that the heroin probably had nothing to do with her problem. They didn't want the sample."

"So what happened to it?" Langston asked.

"We incinerated it."

"Thank you, Officer," Langston said calmly. Then, with an exasperated sigh, he slammed down the receiver. Two days of phone calls for nothing. Suddenly, Langston realized that there was utter silence in the clinic conference room. Residents, students, and even several colleagues were staring at him. Only then did he realize how obvious his frustration had been.

Langston had barely had time to recover from his embarrassment when the phone rang again. It was another law officer on the line, this time from the county crime lab. The caller, one Jim Norris, was extremely interested in the other "heroin" samples Langston had gathered. Norris got straight to the point.

"Dr. Langston, we need the samples of heroin that you seized from the patients for a legal investigation we are conducting."

Langston was taken aback. He hadn't "seized" any samples—he wasn't a policeman—he had asked for them. And he wasn't using them to put someone in jail, but to solve a medical mystery. Why couldn't the crime lab find their own samples? To make it worse, he had just lost a third and critical sample.

"I'd like to cooperate, Mr. Norris," Langston said, "but we need these samples for our research. We don't know what's wrong with our addicts and our only hope is to discover what caused them to freeze up. The answer is in the heroin, and we only have a small amount."

There was a long pause on the other end of the line. "Dr. Langston," Jim Norris began softly, "you don't understand. You can either give us the samples, or I'm coming over there with some of my men and we will take the samples."

Langston felt a chill. This sounded serious.

"In any case," Norris continued, "it's probably not heroin at all." Then he began to tell Langston about the world of designer drugs, about the man they thought might have made this poison which had destroyed the lives of the six addicts and perhaps many more besides, and how the law could not touch him.

Langston now realized he was in no position to negotiate, but decided to try for a compromise. "I understand your situation, Mr. Norris, but I hope you also understand mine," he began. "How about splitting the samples? That way you can conduct your analysis for legal purposes, and we can continue the medical investigation to see if there is any way we can help these people."

Norris agreed. Langston kept half of the bad heroin, and the rest went to the San Jose crime lab for analysis by Halle Weingarten.

Halle Weingarten's careful analysis confirmed that these

samples, like the one from Mason's residence, were from the meperidine family, but beyond that she could not identify the compound. But something stirred deep in her memory. The drug, the formula, the addicts' strange condition—she had read or heard something similar before. That Saturday morning, while she sipped coffee over her morning paper, it clicked. She remembered reading an article in a periodical called the *Journal of Psychiatry Research* some years before about a remarkably similar case, where the victim had used meperidine lookalikes and contracted a Parkinson's-like condition. When Langston called the next Monday morning to ask if she had made any progress in identifying the samples, she suggested that he read the article without delay.

3

The Forgotten Case

Langston had never heard of the *Journal of Psychiatry Research.* Neither had the librarian of Valley Medical Center. But a little research showed that Stanford University's Lane Medical Library held a copy and that the article in question was in the very first edition of this journal in 1979. Langston asked Phil Ballard if he would stop by the Stanford medical library and make a copy. It was on Langston's desk when he arrived at work the next morning. He was immediately gripped by what he read. It told the story of a young college student named Barry Kidston who had used a chemistry set—a present from his parents—to make his own drugs. Barry Kidston had acquired his drug habit overseas when his father, a diplomat, had been posted to India. When they returned to the United States to live in conservative Bethesda, Maryland, Barry reasoned that it would be easier and cheaper to make his own narcotics than to buy them on the street.

So he went to the library and studied the chemical literature to find out how.

In the summer of 1976, after considerable research, he decided to make a compound which had been originally synthesized in 1947 by a Hoffman–La Roche chemist, Dr. Albert Ziering. It was called 1-methyl-4-phenyl-propionoxy-piperidine, or MPPP for short. Barry realized that Ziering's formula described how to make a close analog of meperidine (better known as Demerol). Through what was in effect a designer-drug maker's trick, Ziering had made a small change in the meperidine molecule, making it some five times as potent. Barry reasoned that when injected intravenously it would, like many painkillers, give a heroin-like high.

For several months Barry successfully made MPPP and used it intravenously. One day in November, however, he hurried a batch, and soon after injecting it into his arm he knew that something had gone terribly wrong. The substance burned like it was on fire. Within three days he froze up, became immobile and unable to speak. His parents were terrified and took him to a hospital, where he was initially diagnosed as having catatonic schizophrenia. Since catatonia is considered a major psychosis, the physician in charge made the unfortunate choice of treating him with one of the drugs that can cause parkinsonism, in this case the neuroleptic haloperidol. Not surprisingly, Barry failed to improve. Next he was subjected to electroconvulsive therapy.

When these "therapeutic maneuvers" failed to "bring him out of it," a neurologist, Dr. Ramon B. Jenkins, was called in. Jenkins examined Barry fully and realized that his symptoms—a marked slowing of movement, severe rigidity, tremor, flat facial expression, and difficulty speaking—added up to parkinsonism. Accordingly, Dr. Jenkins treated Barry with L-dopa, which effectively and dramatically unfroze him.

Because of the unusual nature of his case, especially the fact that he had been injecting homemade narcotics, Dr.

Jenkins felt that Barry Kidston needed to be studied in more detail. A few miles away from where Barry lived in Bethesda was the campus of the National Institutes of Health—the world center of medical research. So Barry was referred to NIH's Clinical Center, where a powerful interdisciplinary team of neurologists, psychiatrists, and chemists would study him.

Every year some ninety-five hundred patients come to NIH's Clinical Center, an imposing fourteen-story building that dominates the NIH skyline. This unique research hospital and laboratory complex was designed to bring basic research close to the bedside. Patients like Barry Kidston receive first-class care while NIH scientists get an opportunity to learn more about the diseases that affect them. The scientists studying Barry were affiliated with the National Institute of Mental Health. They agreed that there were a number of possible explanations for Barry's condition. Parkinsonism did occasionally afflict the young—so-called juvenile parkinsonism. But this was extremely rare and certainly did not come on quickly in days. It was also possible that his condition resulted from inappropriate use of tranquilizers that he may have been abusing on the side. But the team's working hypothesis was that Barry had unknowingly made and consumed a chemical which had destroyed the part of his brain important for the coordination of movement.

The scientists were not slow to see the implications of Barry's tragedy for medical research. If Barry had indeed hit on a powerful neurotoxin that selectively destroyed brain cells, that toxin might be immensely useful to science. The scientific study of the major degenerative diseases of the aging brain—Parkinson's and Alzheimer's—had made little progress, for one main reason: there was no good animal model in which to study these diseases. For reasons that are not well understood, animals do not seem to get neurodegenerative diseases like Parkinson's and Alzheimer's. It might be that animals destined to get such diseases do not survive in the wild, because being one step slower means becoming a predator's dinner. Or, for reasons unknown, it may simply be that humans are the only species to get such subtle diseases of aging. Scientists had tried for decades to "model" neurological conditions like Parkinson's in the laboratory, by injecting toxins into regions of animals' brains and even damaging parts of the brain with surgical instruments, but without much success. Scientists had yet to find a toxin that could enter the body of an animal, pass into its brain, and selectively destroy certain cells, thereby producing a complex of signs and symptoms indistinguishable from Parkinson's disease. Perhaps Barry Kidston had discovered such a toxin.

The team of scientists studying Barry was superbly positioned to find out. Not only did they have the resources of the NIH, but Barry Kidston was there to help them. This intelligent young man reconstructed careful notes of his synthesis, including the shortcuts he had taken, and allowed them to take his equipment to try to reconstruct what he had done.

Chemist Dr. Sanford Markey had the job of investigating Barry's chemistry. After talking to Barry, he had gone to the library to read everything he could find both about meperidine synthesis and about drug-induced parkinsonism. Markey was impressed that Barry Kidston had even attempted to synthesize MPPP. The synthesis not only involved some extremely volatile chemicals that would burst into flames if they came into contact with air, but contained multiple steps, all of which needed to be done perfectly. If Barry had used too much acid or too high a temperature, for example, then the synthesis would not work as intended and other side products would be produced as well. If Barry had made unwanted side products, it was possible that these were responsible for his parkinsonism.

Markey wasn't sure how to proceed. He knew how to do the synthesis correctly and make MPPP. What he needed to do was carry out the synthesis incorrectly, reproducing Barry's sloppy chemistry. As Markey set about becoming a "sloppy chemist," he stumbled on a better approach. While setting up Barry's equipment, he discovered that Barry's mother had not cleaned out his glassware completely; there was a tiny amount of residual chemical left in the desiccator grease from Barry's last synthesis. Markey realized that this "dirt" might hold the key to what happened. He carefully redissolved this residue and analyzed it in a mass spectrometer. The results showed that indeed Barry had not made pure MPPP. In addition, he had produced two other compounds. There was some 1-methyl-4-hydroxy-4-phenylpiperidine left over from the first step of the synthesis, and a compound called 1-methyl-4-phenyl-1,2,3,6-tetrahydropyridine, or MPTP. Nothing was known about the effects of these side products.

Markey took the next obvious step: injecting the various compounds into animals to see if he could reproduce the Parkinson's-like state. The lab animal of choice for most biomedical researchers is rats. They are plentiful, cheap, and don't seem to evoke a strong sympathetic response from animal lovers. Markey first injected pure MPPP into the rats and waited.

The effects were dramatic. Within minutes the rats froze up, becoming stiff like boards; some even developed a tremor. It looked convincing, but drugs with opiate-like qualities are known to produce a temporary catatonic-like state in rodents. In fact, this phenomenon is so striking that scientists have jokingly nicknamed the condition "ratatonia." And sure enough, within hours the rats had returned to normal. This experiment was carried out again and again, and each time the results were the same, "ratatonia" followed by recovery. Markey then injected a mixture of the MPPP and the side products into rats. This time nothing happened, so Markey repeated the experiment. But still there was no effect.

Over the next few weeks, Markey tried everything. He increased the dose greatly. He altered the compounds with various chemical substitutes. He tried blocking the liver's normal metabolism (reasoning that the liver was detoxifying the drug). Nothing worked; the compounds which Barry had synthesized had no permanent effect on rats. The scientific investigation stalled. Doubts set in. Perhaps they had expected too much. Perhaps Barry's parkinsonism had nothing to do with his sloppy chemistry. There were other possible explanations. Barry's medical history was rather unusual: he had been bitten by a rabid dog; he had lived abroad for several years and had contracted several exotic diseases; he had abused a whole host of drugs. He was hardly your run-of-the-mill patient. The question the team pondered was this: Was Barry's condition the result of his synthesis in his basement, or was it the interaction of this chemical with something that had happened previously?

If the research team was floundering, so too was Barry Kidston. His doctors had eventually taken him off L-dopa because he was abusing it. L-dopa is a psychoactive drug and some patients feel a rush when it kicks in. Barry, like many drug abusers, had what physicians call "drug abuser's personality." As L-dopa made him feel better, he began to crave that initial rush, and he began to exceed his prescribed dose to get it. L-dopa is a very powerful drug and patients who exceed their prescribed dose sometimes go downhill fast because of increasing side effects. The team switched him to a second drug, called bromocriptine, which had less of a "rush." But Barry continued to abuse other drugs, including codeine and cocaine. In September 1978, after eighteen months of treatment, he became very depressed. One day he came to the campus of NIH, sat down under a tree, took an overdose of cocaine, and died.

Barry's family agreed to let NIH do an autopsy on Barry's brain. They knew how unique Barry's illness was and found some comfort in the thought that his tragic death might help others. The team wanted to do two kinds of studies on the brain: conventional pathology, where sections of the brain would be examined under a light microscope to see whether the same areas of brain involved in Parkinson's disease had been affected; and chemical studies to investigate what chemical damage had been done to the brain. The brain was divided and half of it was stored in a freezer for the chemical studies. The other half was sectioned for microscopic examination. The results were dramatic.

When the brains of patients who died from Parkinson's disease are examined, highly characteristic and recognizable changes are seen. There is extensive cell loss in the area of the brain stem called the substantia nigra. The "black stuff" is literally gone. And so it was with Barry. In the words of the principal author of the *Journal of Psychiatry Research* paper, psychiatrist Glen Davis, ". . . the same areas of the brain were damaged . . . if the (brain) slides had been presented blindly to another neuropathologist . . . they would have called it Parkinson's disease."

Langston could not believe what he was reading. Six years before, a team at NIH had stumbled on a case remarkably like George Carillo. They too had speculated that something in an injected drug had induced parkinsonism—not only the symptoms, but also the pathology. In addition, they had realized something that Langston had not really fully appreciated. If the active toxin was identified and used to reproduce the Parkinson's-like state in animals, it would be a discovery of extraordinary importance. It had the potential to revolutionize an area of medical research. If NIH knew all this, why hadn't they exploited this opportunity? Why was this case languishing unread in an obscure journal?

Langston tracked down Glen Davis and phoned him to get some background information. Davis was now working in Cleveland and was fascinated by what Langston told him of the California addicts. Why, Langston asked, had they ended up publishing in *Psychiatry Research*? Davis responded: "You don't know the half of it. That paper very nearly wasn't published at all." He then began to tell Langston the part of the story which didn't find its way into print, but which reveals much about how medical science operates at the NIH.

Davis had arrived at NIH in 1975, one of dozens of scientists and physicians doing fellowships at the mecca of biomedical research. The National Institutes of Health have no equal anywhere in the world. Situated in Bethesda, about ten miles from Washington, D.C., the 300-acre campus is the nerve center of the world's biomedical research. Through its twenty-four separate institutes, centers, and divisions, it supervises and supports research into diseases from cystic fibrosis to the common cold. Each year it considers some thirty-six thousand research proposals and training applications.

Eighty percent of NIH's $10 billion budget helps fund basic and clinical research at universities around the United States. The bulk of the rest supports so-called intramural research on the Bethesda campus. At any one time some two thousand projects are going on in the seventy-odd buildings on the superbly equipped NIH campus. In addition to the many resident researchers and physicians, thousands of doctors and scientists pass through NIH at some time in their career. Usually, they go for at least two years to study an advanced specialty at one of the institutes: oncology at the National Cancer Institute; cardiology at the National Heart, Lung and Blood Institute; psychiatry at the National Institute for Mental Health; and so on. Glen Davis had gone for a training fellowship in psychiatry. Adrian Williams, one of his collaborators on the Kidston case, was doing one in neurology. The fellowship system works well for the most part, but it is all but impossible for a team of researchers to stay working on a project together for very long. After they had followed the Kidston case for nearly two years, the fellowships of several of the team—including Glen Davis—came to an end, and they left the D.C. area to take up jobs elsewhere.

According to Davis, not everyone on the team (the other members, in addition to Adrian Williams, were Sanford Markey, Michael Ebert, Eric Caine, Cheryl Reichart, and Irwin Kopin) felt as strongly as he did about the Kidston case. It was argued that one shouldn't base too much on a single case. It was pointed out that the animal experiments had not so far yielded anything substantial or, for that matter, even a serious lead. As there was plenty of other research going on, some, like Sanford Markey, wanted to hold off further investigation of the Kidston toxin until there was experimental validation in laboratory animals.

Glen Davis, however, felt very strongly that their preliminary results should be published. Against a background of apathy, he bludgeoned his colleagues into writing a scientific paper about Barry Kidston entitled "Chronic parkinsonism secondary to intravenous injection of meperidine analogues," which he promptly sent off to the world's most prestigious medical journal, *The New England Journal of Medicine.* Four months later, it came back with a rejection which said that *The New England Journal of Medicine* did not in general publish accounts of single cases. Davis was disappointed but not terribly surprised—some of the most famous *New England Journal of Medicine* articles involved thousands of cases followed over years, sometimes even decades. Undaunted, he sent the manuscript to the *Journal for the American Medical Association,* hoping for a more sympathetic response. The manuscript came back with a provisional acceptance, but asked that the number of authors be reduced from the seven named scientists. *JAMA* had instituted a policy aimed at limiting the number of authors jointly submitting papers. This trend, which resulted from some papers having as many as twenty authors, was in their view getting out of hand. They insisted that the Kidston paper, which was a short report of five pages, be resubmitted with no more than six authors.

Davis and his colleagues found this unacceptable. As they saw it, every member of the team had made a contribution and deserved a share of the credit. So, rather than not publish at all, Davis sent the paper to the newly created *Journal of Psychiatry Research,* one of the editors of which, Frederick K. Goodwin, was a senior NIMH scientist, and another, Irv Kopin himself. The paper was accepted and put in the first edition of the journal. Unfortunately, as few scientists knew of this fledgling journal's existence, the article languished unread for the next three years.

It was quite a story. Davis explained that he no longer had any research interest in parkinsonism. "If you want to know more about the chemistry and the animal experiments, you should talk to Sandy Markey. As far as I know, he is still at NIMH."

After Bill Langston finished talking with Glen Davis, he sat for a long time just holding the paper in his hands. The description of Kidston's symptoms was virtually identical to what he had seen in George, Juanita, and the others. The pathology had shown that the substantia nigra had been obliterated, just as it is in Parkinson's disease. The article even gave the details of the chemical synthesis, and the formulas of the side products which would result from using too much heat or acid. Surely, here was the answer to the mystery of the six addicts.

Of all the chemists that Langston had spoken to about the unknown powder, Ian Irwin, the director of Stanford's Drug Assay Lab, had seemed the most knowledgeable and impressive. Langston picked up the phone, called Irwin, and read him the formulas in the article. Irwin quickly calculated in his head the atomic weights of the compounds that Langston was describing, and replied without hesitation: "Well, if it's any of those, it would have to be the by-product 1-methyl-4-phenyl-1,2,3,6-tetrahydropyridine, or MPTP. The atomic weight matches exactly, and the fragmentation pattern is what I would expect with that compound."

Irwin made some inquiries and discovered that it was actually possible to buy MPTP commercially from the Aldrich Chemical Company. It was a minor compound, so minor that it hadn't been characterized for toxicity or listed in the national toxicological database of chemical compounds. Not much was known about MPTP, but Irwin could see how it might have some use as a chemical intermediate. Chemists used chemical compounds like blocks of Legos. A chemist could laboriously assemble a complex chemical compound from individual elements such as carbon and oxygen (individual Legos), or save a lot of time by purchasing preassembled blocks for the construction of even more complicated structures. Over the years, a handful of chemists had ordered MPTP from Aldrich to save some time.

By comparing the molecular fingerprints of the California samples to Aldrich Chemical MPTP, Irwin was able to prove beyond doubt that the chemical was present in significant amounts in the California designer heroin, and that in one sample it made up 97 percent of the powder. Since it was unlikely that the 3 percent MPPP in the sample was to blame, Langston and Irwin now had at least strong circumstantial evidence that the chemical which had caused the addicts to freeze was MPTP, and that this also was the culprit in the Kidston case.

If it was MPTP, why had NIH missed it? Their article had mentioned injecting MPPP into rats. Why hadn't they tried MPTP as well? Langston decided to call Sandy Markey at NIH.

The conversation was pleasant enough. Langston gave a detailed description of the California cases, and how their symptoms seemed to match Barry Kidston's. Then the conversation moved on to the toxicology.

"Dr. Markey, it says here in your article that you tested animals with MPPP—rats, I believe? Did you test the other by-product?"

Markey paused, thinking back to his experiments in the late seventies. "We injected pure MPPP and we were able to produce temporary catatonia, but just about anything does this in rats. As I recall, we then injected a mixture of the by-products as well into the animals, but nothing happened. I'd have to check my notes."

Then Langston asked, almost casually, "Did you try the MPTP, the second by-product, on its own?"

Silence. After what seemed like an eternity, Markey answered. "You know I think that's the only one we didn't specifically try out on its own."

Langston was exhilarated. MPTP, he felt sure, was the neurotoxin. He would bet money that if pure MPTP was injected into a rat, that rat would develop parkinsonism. In just three weeks his professional life had completely turned around—or at least it felt that way. Now he was a driven man, he had a purpose. A circle had been closed, linking illicit chemists, frozen drug abusers, and scientific research teams working in different places and at different times. This time the significance of the sloppy meperidine synthesis to Parkinson's would not be missed.

But from now on he would have to share the case with NIH. They would marshal their enormous resources and expertise and try to solve this mystery once and for all.

4

Brain Damage

The brain is an organ of awesome complexity, but it has one big flaw. Once nerve cells in the brain die, they are gone forever. Brain cells cannot replicate after injury to replace nonfunctioning or dead nerve cells. Unlike skin after a wound, they can never grow back. We are, therefore, all born with a full complement of neurons; after birth our brains will never produce another nerve cell.

Because of this vulnerability, the brain has evolved several lines of defense against losing nerve cells in the first place. One of these is physical. To cushion it from outside blows, the brain floats in a special liquid—cerebrospinal fluid—which also surrounds the spinal cord. The brain and spinal cord are also sheathed in three layers of membranes and encased in bone.

The second line of defense is against chemical damage. The blood vessels in the brain are not like the blood vessels in the rest of the body. They are constructed so tightly that only the smallest molecules can pass through their walls. The vessels thus act as a protective filter against most chemicals or other substances that might be damaging to the brain. Most dangerous molecules can't get through this filter, which is known as the blood-brain barrier.

With these kinds of defenses, how had MPTP—if indeed it was MPTP—managed to slip into George's brain and wreck his life?

Pondering this question, Langston reread for the tenth time the *Journal of Psychiatry Research* article. In the footnotes the authors had cited a number of old articles which had inspired Barry Kidston. As the Valley Medical Center library didn't have these articles, Langston drove to Stanford to use the library system there. He consulted chemical abstracts and looked up all literature relating to the synthesis of MPPP and made a list. Then he walked downstairs to the electronically controlled stacks to begin his search.

Langston located the stack which contained the *Journal of Organic Chemistry* for 1947. He found the right volume and began searching for the article "Piperidine derivatives: Part III" on page 894. He flipped through, looking for the right page: 892, 893, 899, 910 . . . Langston thought he had made a mistake. He checked again. There was no page 894 or page 895. There were no pages 896, 897, or 898. All the pages relating to the article were missing. They had been razor-bladed out. Annoyed, Langston looked for the next reference, in the *Journal of Pharmacology and Experimental Therapeutics*. He found volume 91 and began searching for the first page of the article, "Piperidine derivatives with morphine-like activity." Again, the article was gone. His next reference was an article from the same journal, published in 1948, "Pharmacologic studies on analgesic piperidine derivatives." Yet again, the article had been torn out of the volume. Langston's imagination began to go wild. This must have been the very library where the California designer-drug chemist had done his research. Perhaps he had stumbled onto the trail of the person whose acts had crippled Connie. If so, why hadn't the chemist used the photocopying machines? Perhaps he wanted to limit competition from other designer-drug makers?

Returning to VMC, Langston looked in on George. L-dopa had completely transformed him. He could move and talk virtually as well as before the tragedy. He seemed to enjoy the attention he was getting and liked talking to the media. Over the short time they had known him, Langston and Ballard had discovered George to be an extraordinary character. Small, thin, wiry, and very tough, George had been involved in many criminal enterprises, from petty theft to narcotics and prostitution. He had served time at some of California's most famous jails. He had a wife and children, but they had turned him out years before.

For all this, he had an artistic streak within him which might have flowered if his upbringing had been different. He liked to paint and write poetry. George, who had grown up in Silicon Valley, had never stuck with anything. He had wanted to be a singer but had ended up dealing drugs. He had tried marriage and a family and ended up as a petty criminal. He had never held a steady job.

In a few days George would be returned to jail to serve out his sentence, but Langston would make sure he came to VMC for regular follow-ups.

Juanita had responded so well to L-dopa therapy that Langston had discharged her to outpatient care. Langston had found her to be a sweet, caring person, and like George's, her story had drawn him in. Born on January 2, 1952, in Fontana, near Los Angeles, Juanita was one of eight children. Her family had moved to Gilroy in the agricultural heartland of California and she attended school in Watsonville until the eighth grade. As a child growing up, her favorite activity was going to church and her ambition was to become a nun. But things didn't turn out that way. She began taking drugs when she was twelve years old and gave birth to an illegitimate son in her teens. Thanks to a warm, supportive family, Juanita had somehow stumbled through life before taking the bad heroin. Langston intended to keep seeing her on a regular basis to monitor how the L-dopa was working.

The Silvey brothers had also responded very well to L-dopa, and were being treated in Watsonville by their physician, Dr. Murphy, and the consulting neurologist, Jim Tetrud. David and Bill Silvey had grown up in Watsonville and had been abusing drugs since their early teens. Apart from occasional work as laborers and truck drivers, the brothers had not found steady work and were in constant trouble with the law. They were well known to the Watsonville police and had long criminal records. Thanks to Drs. Tetrud and Langston (and, of course, the L-dopa) they could both move again. So far, they had kept out of trouble and had not gone back to abusing drugs, but Langston suspected it would just be a matter of time before they did.

As far as Langston knew, Toby Govea was the first person to have taken the bad heroin and developed symptoms. He reported first taking the heroin in late April, yet because his condition was not correctly diagnosed, it was August before he was put on L-dopa. Toby responded very well to a mixture of L-dopa and bromocriptine (a drug which prevents the L-dopa wearing off too suddenly) taken every three hours. His symptoms virtually disappeared, although they returned immediately if the medication was lowered or stopped.

The patient that Langston was most worried about was Connie Sainz. Initially she had been put on three 10/100 Sinemet (carbidopa combined with L-dopa tablets), but getting little effect, Langston had doubled the dose. Connie had regained some facial expression and was able to talk a little, but it seemed the dose was still inadequate.

Eventually, Langston had increased Connie's dose to six 25/250-milligram tablets, combined with bromocriptine to make the L-dopa last longer. This time it worked. Connie's parkinsonism was reversed and she was able to move and talk.

Connie and the others were walking and talking, thanks to L-dopa, a remarkable drug that had transformed the lives of Parkinson's disease sufferers. Before L-dopa, the average life span of patients with Parkinson's disease was seven to ten years. As a medical student Langston had read the astonishing story of brilliant medical research which led to L-dopa's discovery. In the 1950s, before Parkinson's had even been linked to dopamine, the Swedish scientist Arvid Carlsson had carried out a series of ingenious experiments in which he gave rabbits reserpine—a drug which for a few hours causes them to become very slow, apathetic, and nearly paralyzed; even their ears wilt. Carlsson theorized that their "motor problems" resulted from a chemical imbalance in a region of the brain called the striatum and wondered if the rabbits' condition could be correlated with a depletion of the common neurotransmitter dopamine.

To test his theory, Carlsson had set out to reverse the animals' parkinsonism. He injected a substance called dopa, which passed into the rabbits' brain, where it could be converted into dopamine. The results were dramatic: dopa restored dopamine levels to normal in the rabbit striatum, and completely reversed the symptoms of immobility. Carlsson was aware that reserpine was capable of causing a Parkinson's-like condition in humans. Putting all of this together in a now famous lecture at the National Institutes of Health in 1958 as part of the First International Symposium on Catecholamine Metabolism (later published in 1959), Carlsson suggested that a dopamine deficiency might be the neurochemical basis of Parkinson's disease. It was one of those flashes of insight that characterize great discoveries in modern science. Yet Carlsson's proposal that dopamine was important for motor behavior was at first greeted with great skepticism, even downright rejection, from several leading authorities.

However, Carlsson's work did not go without notice. It attracted the attention of a young Austrian scientist, Oleh Hornykiewicz. Following Carlsson's suggestion that a shortage of dopamine might be the critical factor in Parkinson's disease, Hornykiewicz obtained human autopsied brains for analysis. Hornykiewicz and a coworker, Herbert

Ehringer, discovered that the brains of deceased patients with advanced Parkinson's disease had virtually no dopamine in the striatum. Then, together with the Austrian neuroscientist and physician Walther Birkmayer, Hornykiewicz continued this work, reporting that not only was the striatum depleted of dopamine, but also the substantia nigra. Hornykiewicz and Birkmayer then made the obvious suggestion of trying to increase the supply of dopamine to a patient with advanced Parkinson's disease. The question was, would dopa do for people what it had done for rabbits?

As Birkmayer and Hornykiewicz reported in a paper in 1961, after injecting dopa the results were dramatic. "The effect . . . was, in short, a complete abolition or substantial reduction of akinesia. Bedridden patients who were unable to sit up, patients who could not stand up from a sitting position, and patients who, when standing, could not start walking, performed all these activities with ease . . . they walked around with normal associated movements and they could even run and jump . . . This dopa effect reached its peak within two to three hours and lasted, in diminishing intensity, for 24 hours."

Birkmayer and Hornykiewicz's findings were independently replicated by Andre Barbeau in Canada, who for the first time gave the drug orally. But there were problems. Barbeau ran out of money (dopa was at the time very expensive) and couldn't continue his experiments. Birkmayer and Hornykiewicz became pessimistic about dopa as a treatment for Parkinson's disease because of the difficulties in administering large amounts of dopa without producing nausea, vomiting, and hypotension in the patients. It was not until 1968, when George Cotzias, a scientist and physician at the Medical Research Center, Brookhaven National Laboratory, in New York, reported dramatic effects with oral dopa in *The New England Journal of Medicine,* that the drug finally became a therapeutic reality. Cotzias had succeeded partly by using a pure "levo-" ("L-") dopa form of dopa. (Just as our hands are identical but not superimposable, molecules of identical structure can be left- or right-sided too; this is denoted by and "L" (levo) for left, and "D" (dextro) for right). All earlier studies had been done with a mixture of the two forms. Also, Cotzias used a graduated-dose regime: starting with very low doses of L-dopa, he slowly increased the amount given until patients could tolerate large doses without adverse effects—an average dose of around 5800 milligrams of L-dopa. Prior to that, most groups had been working with 50 to 100 milligram doses.

By the early 1970s L-dopa was in widespread use, hailed as a miracle drug—the answer to Parkinson's disease. Patients who were severely crippled with Parkinson's could, after L-dopa, move and talk almost normally. It was as if their disease vanished for as long as the dopamine remained in their brains—typically about three or four hours. So dramatic was the effect of L-dopa that the clinical trials into its efficacy were stopped. It clearly worked. Basic research into Parkinson's disease waned. Surely, most physicians agreed, L-dopa was the answer.

The first glimmerings that all was not well came when some physicians (Andre Barbeau was the first) began to notice that after several years of using L-dopa successfully, many patients experienced strange side effects. While initially these reports were dismissed, by the mid 1970s physicians agreed that L-dopa therapy had its problems. First, with time many patients started developing excessive movements known as dyskinesias, which could become so severe that physicians were forced to decrease the dose of L-dopa, even though it was needed more than ever because of advancing disease. Second, after a patient used L-dopa for a few years, the medication was effective for less and less time. To sustain the miracle, L-dopa had to be taken more and more frequently—in some cases as often as every forty-five minutes. Moreover, with time patients developed a tolerance to the medication, needing more and more L-dopa per dose to achieve any satisfactory therapeutic effect.

The most devastating problem, however, concerned rapid fluctuations—so-called on-off effects—in which the medicine's power to combat Parkinson's disease seemed to vanish suddenly, leaving a patient frozen or stuck midway through a movement or conversation. Minutes or sometimes hours later the medication could just as inexplicably switch back on. These rapid fluctuations seemed to occur randomly, making it literally impossible for some patients to undertake any planned activities.

Some patients also experienced confusion, agitation, paranoia, and hallucinations with long-term L-dopa therapy, which were even more disruptive and at times more disabling than the illness itself.

Many strategies were attempted to avoid these side effects. Some physicians favored putting the most severely affected on periodic "drug holidays," where for several weeks patients would be taken off all L-dopa, in an attempt to restore some of the medication's original effectiveness and decrease the sensitivity to side effects. Other doctors juggled cocktails of drugs, combining L-dopa with medications which modified its action.

By 1980, it was clear that the initial optimism over L-dopa had been misplaced. It was not a cure after all, nor was it the final answer to Parkinson's disease. Patients certainly lived longer in 1980 than in 1960, but their long-term prospects remained bleak. More than a decade of experience of using the drug showed that L-dopa did not stop the progression of the disease. Even as patients took L-dopa, their dopamine-making neurons continued to die.

Langston was not sure if Connie, George, and the other addicts would succumb to the same kind of side effects as Parkinson's disease sufferers and if so when? In regular Parkinson's disease patients, the more troublesome side effects usually don't show up for five or more years. For the time being at least, L-dopa was working its miracle with the frozen addicts.

In George, Langston had found the quintessential case for his neurobehavior unit. George's paralysis, his facial masking, his drooling, his stare, all resulted from a lesion—probably a very tiny lesion—in his brain. Something in the drug, probably MPTP, had killed some of George's brain cells, thus destroying his capacity for voluntary movement. L-dopa had restored that capacity, at least for the time being.

Many times since the tragedy, Langston had tried to imagine what had been going on under George's skull. When George had lain frozen in the hospital bed, his nerves still carried information to his brain from the outside world: the image of the doctors, the sound of their voices discussing him, the searing odor of the smelling salts they put under his nose. His brain interpreted this information in the context of past experiences and then sent out commands which passed down nerves to different parts of his body. He tried to call out. He tried to flinch. He tried to hit one of them. But his body didn't respond.

While only weighing three pounds, George's brain, like all adult brains, had some 100 billion cells, called neurons, each neuron being able to connect with thousands of other neurons. The upper portion of a brain—the cerebral hemispheres—can be thought of as the cap of a mushroom and the rest of the brain, as its stem. This so-called brain stem connects the brain to the spinal cord. This region of the brain also regulates all of the normal body functions: heart rate, blood pressure, breathing. Cranial nerves exit from the brain stem that control muscles in the face, tongue, eyes, ears, and throat, and return sensations from these parts back to the brain. In the very top of the brain stem is the substantia nigra, the tiny region of cells which normally makes dopamine, but which in George's case (like Barry Kidston's) had in all probability been destroyed by the bad heroin.

Under the microscope, neurons look quite different from other cells in the body: they have extensions. Sprouting out from the cell body are masses of short, tiny branches, called dendrites, which connect with other cells. These dendrites receive incoming messages. Most nerve cells in the brain also have one long fiber which can extend considerable distances—up to several meters in the case of certain nerve fibers that run from the brain to the spinal cord. This fiber,

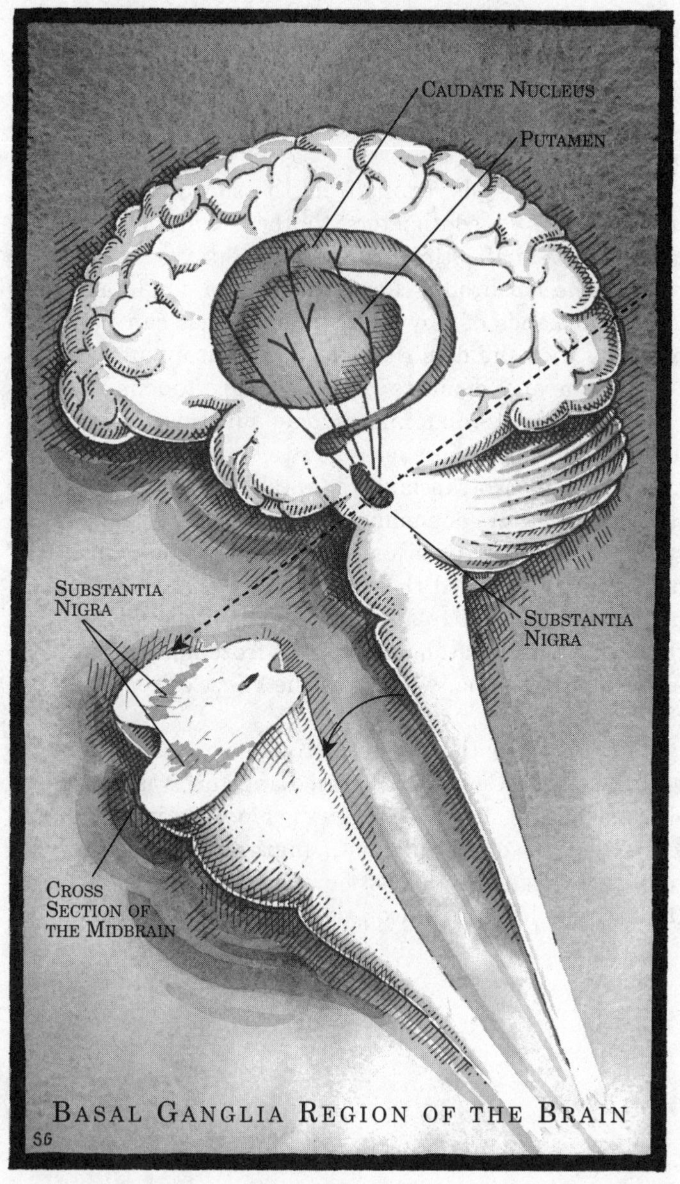

Axons from the substantia nigra, a major source of dopamine, extend to the striatum, a critical nerve junction that incorporates the caudate nucleus and the putamen. This dopamine pathway is essential for normal movement; it enables the thought of moving to be converted into the act of moving. When the cells of the substantia nigra begin to die, as in Parkinson's disease, dopamine production drops and movement becomes increasingly difficult to initiate and carry out.

the axon, can branch many times, and carries outgoing electrical impulses to other cells, connecting with the dendrites of those cells.

In healthy people, axons from the substantia nigra extend about two centimeters to the striatum, a critical nerve junction affecting movement, positioned deep in the base of the mushroom cap. The end of a single axon has many thousands of tiny nerve endings that can form connections with the dendrites of other neurons in the striatum. Axon and dendrite don't actually touch, but pass chemical messages at microscopic junctions called synapses. Axons pass these messages to dendrites by releasing chemicals—neurotransmitters (in the substantia nigra, the neurotransmitter dopamine)—which cross the synaptic space and bind to the other neurons' membranes. If MPTP had indeed destroyed the neurons in George's substantia nigra, this would explain why no dopamine was getting through to his striatum and why he froze up. It would also explain why L-dopa, which has the effect of replacing the missing dopamine, worked.

Based on the pathology of the Kidston case and his observations of the California cases, Langston was convinced that MPTP had slipped through the blood-brain barrier and killed the cells of the substantia nigra. It had also occurred to him that figuring out just how this happened might turn the entire field of Parkinson's disease upside down.

5

Research Wars

August 1982

After Langston's call, Dr. Irwin J. Kopin, chief of the Laboratory of Clinical Science for the National Institute of Mental Health, and a coauthor of the *Journal of Psychiatry Research* paper, and Sandy Markey began an intensive review of NIH's investigation into Barry Kidston.

Kopin was a rising star at NIH. It was strongly rumored that within a few months he would become scientific director of the National Institute of Neurological Diseases—one of the most powerful jobs in neuroscience. In this position he would be a key figure in brain research, overseeing dozens of research projects.

The NIH is a highly political institution. To succeed there, a person must be not only a good scientist but also an accomplished student of power. Kopin understood power and sensed the importance of what had happened in California. Some unknown clinicians at a county hospital in San Jose had apparently stumbled on an epidemic of drug-induced parkinsonism, and the parallels to the Kidston case appeared very strong. If the California cases had injected the same compound as Barry Kidston, then they might be on the verge of a major medical breakthrough.

Markey had meticulously reviewed his experiments with Kopin, how he had injected first the pure MPPP and then the mixture of all the side products into rats and how nothing unexpected had happened. Markey knew that these experiments had to be repeated more systematically. All three compounds—MPPP, the hydroxyphenylpiperidine, and the MPTP—would have to be separately injected in varying doses into rats to see if anything happened. The California cases provided very strong circumstantial evidence that MPTP was the culprit, but that would have to be proved.

With Markey was a neurologist named Stan Burns. In his thirties, Burns was midway through a fellowship at NIMH studying the kinetics of the distribution of dopamine metabolites (chemical breakdown products) in monkeys. Markey had involved Burns not only because he was a trained neurologist, but also because he had worked in the field of drug abuse for several years prior to coming to NIH. In fact Burns's motivation in taking this fellowship had been to learn the latest biochemical methods so that they might be used with drug abusers. While Burns had not been directly involved with the Kidston case, he had followed it with interest. Now he was really excited. These California cases sounded so similar to Barry Kidston's that there *had to be* a connection. There *had to be* a toxin common to all the drug samples that killed cells in the substantia nigra. Burns knew that the discovery of a neurotoxin—a chemical which could reproduce the symptoms of Parkinson's disease in animals—could revolutionize research in Parkinson's disease.

It was agreed that Markey would repeat his lab experiments in rodents, while Burns and Kopin would try to get access to the California patients to see if they really had drug-induced parkinsonism. Kopin called up Langston to tell him that the resources of NIH were at his disposal. He told Langston all about their research programs in Parkinson's disease and how they had built up a unique expertise in this area. "We are here to help, Dr. Langston; please tell us if there is anything we can do. Here at NIH we can carry out many different types of clinical investigation that may not be possible at your hospital. For example, we can study spinal fluid for the breakdown products of dopamine."

Given NIH's expertise, not to mention their experience with the Kidston case, it made sense for them to be involved. But Langston politely declined the offer of collaboration. "At the moment we have everything under control—the patients are doing well on L-dopa and Stanford can, I believe, do any chemical studies we need. But many thanks for calling, Dr. Kopin. If we run into any problems, I will certainly take up your generous offer."

The following day Stan Burns called Langston to make his offer of help. Burns needed to be sure of two things: that these California patients really had parkinsonism, and that the compound they injected was a meperidine analog like the one Barry Kidston had taken. If the California cases had been struck down by the same toxin as Kidston, then Burns was anxious to get his hands on these patients. He felt he was uniquely qualified to take advantage of this opportunity. His background in drug abuse prepared him for dealing with the California addicts, his skills as a neurologist enabled him to diagnose parkinsonism, and the biochemical techniques he was using on monkeys could be applied to some of these patients to see whether levels of metabolites in the cerebrospinal fluid could be correlated with the parkinsonism.

But Langston didn't take up Burns's offer of help, and repeated what he had told Kopin.

Burns was not happy. If Langston wouldn't help him, he would have to find another way. He spent the rest of the afternoon on the telephone calling California. He spoke first with the Silveys' physician, Dr. Sean Murphy, then with Dr. Jim Tetrud, and finally with the Silveys. On the phone, they sounded pretty unhappy. But they agreed to see him.

The next morning, Burns flew to San Francisco, rented a car, and started driving south along Route 101. His car passed through Silicon Valley and San Jose and on into California's agricultural region, where Connie and Toby had grown up. He sped by enormous fields of broccoli, garlic, and lettuces growing in the dry heat. Among the sprinklers, armies of mostly Hispanic workers toiled in the fields picking fruit and vegetables. Many had come from Mexico and Central America, some legally and some illegally. All were prepared to endure the poor working conditions and the low pay.

Remarkably, this rural agricultural region was a major drug-abuse center, a place where PCP, heroin, cocaine, amphetamines—anything that blunted reality—was routinely consumed. Adult workers abused drugs, adolescents abused drugs, families abused drugs together. There was less of the violence that accompanies drug use in the inner cities; rather, drugs were seen as part of a way of life, something that you had to take to get through the day.

After driving for nearly three hours on Route 101, Burns turned right on Route 156 heading toward Watsonville and the coast. The flat, cultivated land gave way to green rolling hills which looked lovely in the soft evening light. Burns had to decide what he would do when he got to Watsonville. In addition to examining the Silvey brothers, he would need to take them off their L-dopa to see if their parkinsonism returned. He also needed to obtain a sample of the material they had injected for Sandy Markey to analyze. Only then would he know for sure whether it was MPTP or MPPP that had produced the neurological damage in David and Bill Silvey.

Because the tests on the Silveys would take several days, he had decided to take them back to NIH with him. He hoped that they and their personal physician, Dr. Murphy, would be agreeable.

Valley Medical Center, August 4, 1982

Bill Langston looked at his watch and saw that it was almost time for lunch. He was sitting in a room with a group of Stanford medical residents who were doing a neurology elective at VMC. Langston was feeling somewhat irritated with this group. They had impeccable academic credentials, but their practical abilities and bedside manners could be abysmal. "All right, Derek, please summarize the case we just saw."

A good-looking and well-groomed man in his mid-twenties stood up and began to talk nervously. "The patient, Mr. Bonner, is a 49-year-old man from Mountain View. He was admitted to the hospital complaining of stiffness and joint pain . . ." As Derek talked, Langston's mind wandered. It was almost three weeks since he had first seen George. They had been three of the most exciting weeks of his life, involving patients, drugs, police, and medical research. Langston kept thinking about the addicts and NIH. He wondered whether he should have acted more positively toward Kopin and Burns. He had not consciously sought to exclude the NIH. As a clinician, his main commitment was to the welfare of his patients, and as far as they were concerned, things were under control. They had responded spectacularly well to the L-dopa. Of course the NIH's interest was understandable, given the Kidston case.

Derek had hit his stride. "On examination, the patient appeared to have stiffness in the upper limbs . . ." As Langston listened, he thought about the implications. The case that Derek was discussing, Mr. Bonner, clearly had early Parkinson's disease. Like hundreds of thousands of others, he would get progressively more and more disabled. In ten to fifteen years' time, the bright, intelligent, middle-aged man he had talked to that morning could well be an invalid. A cure for Parkinson's disease would be a wonderful thing.

Langston sighed. In the meantime, it was his duty to make sure that new doctors at least knew how to diagnose Parkinson's. "So what is your differential diagnosis, Derek?"

Derek paused. "It might be early symptoms of Wilson's disease."

Langston groaned inwardly. As usual, they were looking for zebras rather than horses. "Thank you, Derek." Everyone's eyes were fixed expectantly on him.

Langston decided to give them something to think about. "Before we break today, I want to say a few things. The study of neurology requires a special kind of doctor. He or she has to be especially vigilant to small details. If a neurologist can detect a major neurological disease in its infancy, this can sometimes make an enormous difference to the life of a patient. The patient today, Mr. Bonner, has the early symptoms of Parkinson's disease. Since Derek didn't notice the critical signs, I will quickly go through them for you. What did Mr. Bonner say when we asked him about how he moved around the house?"

Several residents answered at once. " 'It's hard to get up out of a chair.' "

"Yes." Langston continued. "In fact, his words were, 'I just don't seem to be able to get going in the mornings. Getting out of a chair. Getting from place to place. It's like I'm being held back. Once I start moving it's okay.' This is an excellent description of akinesia—the difficulty of initiating movement. This is a cardinal symptom of Parkinson's disease. What are the other classical features of this disease? The patient's movements are slowed, something we call bradykinesia—and the patient tires easily. Mr. Bonner, you remember, complained of this as well. Other cardinal features include tremor at rest and decreased muscle tone. Remember, he complained of stiffness? But there are many other symptoms that may occur in the parkinsonian patient. The voice becomes monotone and softer and softer, a symptom referred to as hypophonia. As the condition progresses, patients are often asked to repeat themselves, and sometimes can't be understood over the phone. Facial expression is typically lost, which is called hypomimia, so much so that patients are often asked why they are depressed, even when they are in good spirits. The patients develop a stooped posture and an abnormal gait, typically walking with very small, shuffling steps, and with a greatly reduced or even totally absent arm swing. Patients cannot keep their balance and often fall forward. One way we test for this is to pull them backwards, throwing them off balance." Langston paused. Now he had their attention. "What else do we notice, when the disease gets more advanced?"

"Freezing," someone replied.

"Yes. Freezing episodes are common, where the patient simply stops, as if turned to stone. The muscles are rigid, which at times causes painful cramping, particularly in the legs. Of course, the most familiar symptom—and one clearly visible in Mr. Bonner—is the tremor. In many cases this is the first symptom to appear. The tremor almost always affects one side more than the other and it typically occurs in the resting state—that is, when the limb is not doing anything. The tremor decreases or stops with movement, typically gets worse with stress, but disappears during sleep."

Langston was on the point of telling them about the addicts, but decided against it. He proceeded to wrap things up forcefully. "Parkinson's disease is one of the great mysteries of modern medicine. What makes it particularly tragic is that the patient is usually mentally normal, with all of his or her intellectual faculties intact and all of his emotions as strong as ever. As the disease progresses, in a way the patient becomes a prisoner in his own body. The kinds of movements most affected in Parkinson's disease are voluntary movements. It is paradoxically the conscious intention to move that is the problem. A severely disabled Parkinson's disease patient who can't intentionally pick up a set of keys on a table may nevertheless be able to catch a ball that is thrown to him by his physician. Catching the ball is a reaction, almost an involuntary movement. Yell 'fire,' and a parkinsonian patient who hasn't moved in years will be the first out of the house, only to become immediately frozen again. We call this 'kinesia paradoxica,' but have no idea why it occurs. If we did, we might be able to take advantage of the phenomenon therapeutically."

He turned to Derek, who looked decidedly uncomfortable. "Finally, Derek, you shouldn't feel too bad about getting this case wrong. Parkinson's disease can be a very difficult disease to diagnose. In London there is a brain bank containing the brains of a large number of Parkinson's disease sufferers, who donated their brain tissue for medical research when they died. One of the most remarkable things about this brain bank is that nearly one out of every four patients who were diagnosed as having had Parkinson's disease during life by experts didn't have the disease at all when their brains were examined. In other words, nearly 25 percent of the patients were wrongly diagnosed. Okay, time for lunch."

Langston was almost out of the ward, heading toward the cafeteria, when a call came through from Dr. Murphy, the Silveys' family physician in Watsonville. Murphy seemed a bit agitated.

"Dr. Langston, did you know that there is a doctor from NIH down here?"

Langston had no inkling of what he was talking about. "No, Dr. Murphy. What's his name?"

"Dr. Burns."

"Stan Burns?"

"Yes. He's been down here trying to locate the Silvey brothers."

Langston couldn't believe what he was hearing. "My God, I'm just stunned."

But Murphy continued, "Well then you probably don't know that he has been to see David and Bill Silvey, and he intends to take them back with him tomorrow to NIMH. I was a little concerned, especially when I heard that Dr. Burns has taken them off their medication."

Langston was speechless. In his view, Burns's actions were unconscionable. David and Bill Silvey were two of the six cases he had been studying. While they weren't technically being treated at VMC, from a scientific standpoint they were patients he was investigating. And from a purely medical standpoint, they were Dr. Murphy's and Dr. Tetrud's patients. Burns therefore had intervened directly between them and their patients, changed their medication, and apparently planned to take them back to the NIH with him for his own scientific ends. And Burns had done all this behind his back. What was going on? The irony of it all was that if Burns had been up-front about his visit, Langston probably would have met him at the airport and arranged for him to examine them. Instead, now he was seething with anger.

In that instant, Langston realized how incredibly naive he had been in refusing the help of NIH. This may have begun as a clinical mystery involving an indigent drug addict in a county hospital, but now it had become something else. Langston had taken NIH's offers of help at face value. Perhaps they had been coded statements, the rough translation being "Thank you, Dr. Langston, for finding these cases. We are taking over now; please move aside."

Langston didn't know what to do. Should he complain? Who should he complain to—to Irwin Kopin at NIMH, to the dean of the Stanford Medical School, to the director of Valley Medical Center? What would he say? But as Langston thought it through, he realized that all might not be lost. Burns might just have miscalculated. By themselves, David and Bill Silvey would not be of much scientific interest to NIMH. To make any kind of evaluation NIMH would need not only the patients but the synthetic heroin they used as well. After locating the Silveys, Burns would probably go to the Watsonville police to ask for samples. The police would tell Burns that the heroin they had found in the Silveys' apartment had been sent to Drs. Ballard and Langston. Sooner or later, Langston reasoned, Burns would show up at VMC trying to get his hands on the powder.

And so it happened. At five o'clock in the afternoon, as Langston and Ballard were doing rounds in the neurobehavior unit, a nurse announced that a Dr. Burns was in the lobby and wanted to talk with Phil Ballard. Langston immediately said, "Let me handle this, Phil."

Outside in the waiting room, Langston could see Stan Burns pacing nervously. It was not hard to guess how Burns had rationalized his behavior. He probably felt that Langston had left him no choice, having turned down NIH's offer of help. Perhaps, Langston speculated, Burns thought this discovery was too important to leave to the Santa Clara County Valley Medical Center.

As Langston entered the waiting room, Burns smiled broadly and put out his hand. "Good afternoon, Dr. Ballard."

But Burns's smile was not returned. "My name isn't Ballard, I am Dr. Langston, and I want to know what the hell is going on."

Burns went pale.

For two hours Langston laid into Burns. What did he imagine he was doing, going behind everyone's back? Did he think it was acceptable to commandeer somebody's patients? Would he like a physician to intervene between him and one of his patients and change their medication unilaterally? Langston could not remember being this angry. But Burns didn't apologize. Try as he might, Langston couldn't get him to see that he had done something wrong. He had never met anyone so impervious to criticism or rational argument. Frustrated, Langston abandoned the argument and left for home.

That evening, as his wife, Lisa, put their infant son to bed, Langston reflected on where his life was going. He was clearly at a turning point. He could simply turn over the cases to NIMH and continue his life the way it had been three weeks before, or he could try to do something with this opportunity fate had thrown him. He had made an important clinical discovery, but perhaps his role was now over. He had no laboratory, no staff, and no money. He had written only one paper in the area of Parkinson's disease and had no reputation in the research community. NIH, on the other hand, had everything. They were the premier biomedical research center in the world. They were set up to do most kinds of laboratory work and any kind of clinical study. If this did turn out to be a significant breakthrough, which could one day help millions, perhaps it was his duty to step aside and let them take over.

Langston called his mentor, Lysia Forno, a neuropathologist at the Veterans Administration hospital in Palo Alto and one of the world's leading specialists in the pathology of Parkinson's disease, to ask her advice. She was candid. She understood why he was angry. He had every right to be upset. But NIMH had a genuine contribution to make. True, Langston's careful clinical work had exposed the phenomenon, but was he really set up to solve the mystery alone? He was a clinician with a full caseload and intensive teaching responsibilities. He had no laboratory experience, let alone a laboratory; no equipment, and no animals for research. With time, a California research team might be put together; she could even do the neuropathology. But currently he had no team. On the other hand, NIH had phenomenal resources and first-class researchers. NIH had studied the Kidston case and had slides of Kidston's brain. The rational, scientific course was to find a way to collaborate with NIH and try to solve the mystery together.

By the next day, Langston had decided that Lysia Forno was right. Langston knew that his own naïveté was partly to blame. Yes, he would agree to let the Silveys go to NIH, and would send them a sample of the contaminated heroin the Silveys ingested. But in return he wanted NIH to keep him informed of the treatment the Silveys received, and also to send Lysia Forno slides of Barry Kidston's brain to review. Langston wrote up a letter of agreement which Stan Burns was to sign and return. The matter was settled.

6

A New Start for Parkinson's Disease

Veterans Administration Hospital, Palo Alto, California, Fall 1982

Lysia Forno adjusted the microscope slide until the image was pin sharp. To most doctors, the image she was seeing would be meaningless. But her expert eyes could see the important details. When patients die of neurological conditions and physicians want to know whether or not they were diagnosed correctly, they send the brain for pathological analysis to specialists like Forno. The brain sample is fixed in formalin and embedded in molten wax. When the wax has solidified, the block is cut into ultrathin slices which are stained and laid on a glass slide, then crossed with another thin slip of glass. Looking at the slide through a light microscope, Forno searched for signs that invariably accompanied a known disorder like Alzheimer's or Parkinson's disease.

The slide, which Stan Burns had sent from NIH as part of the new collaboration, was a section of Barry Kidston's brain. Lysia Forno was struck by two things. First, only one area of the brain was affected, the substantia nigra. Second, the cells in this region were almost completely destroyed. In normal Parkinson's disease, other parts of the brain are damaged as well. Perhaps this told scientists that most if not all of the symptoms of Parkinson's disease resulted from death of substantia nigra cells alone? Forno looked for the other classical hallmark of Parkinson's disease, Lewy bodies. These mysterious round microscopic bodies are pink-staining, with a lighter peripheral halo. Named after the French neuropathologist Frederick H. Lewy, who first described them at the beginning of the century, Lewy bodies are typically seen within degenerating nerve cells. After carefully scrutinizing the slide, Forno found a single pink oval shape which looked like a Lewy body, but wasn't an exact fit. Moreover, in Kidston's brain, this single Lewy-like structure was not in a nerve cell, but in the space between cells. She couldn't say, therefore, that it was precisely the same as in Parkinson's disease. But it was incredibly close. Forno had never seen anything like it before, and she understood why the scientists at NIH were so excited.

Forno had always been fascinated by Parkinson's disease. First described by the English physician James Parkinson in 1817, it was one of the most perplexing conditions in medicine. She thought it ironic that her colleague Bill Langston had been challenged by six mysterious cases; James Parkinson too had been led to his conclusions by following six unusual patients. Like Langston, Parkinson had noticed that his patients possessed an array of neurological symptoms that seemed to represent a distinct disease. Parkinson had carefully noted most of the classical features of the disease with such skill and succinctness that to this day medical students are instructed to read his account as a model of clinical description.

Once Parkinson had described this disease, other physicians began to notice cases and statistics were collected on its prevalence. It was discovered that Parkinson's disease occurred all over the world and showed an increasing prevalence with age, rarely occurring before the age of 40, and becoming more common at least through the seventies and eighties.

One hundred seventy years after James Parkinson's description, doctors had become increasingly adept at diagnosing and analyzing Parkinson's disease, but treating the disorder was challenging, particularly in its late stages. Forno was glad that she wasn't a clinical neurologist, having to help desperately disabled patients with this terrible disease on a daily basis.

Forno admired those like Bill Langston who dealt with patients on the front line of medicine. She wondered if this discovery of drug-related parkinsonism might indeed be very important. If it increased interest in Parkinson's disease research, that would be wonderful. Regardless of how the future collaboration with NIH turned out, she had counseled Bill Langston to try to publish his preliminary clinical findings in a scientific journal. She thought that Bill Langston, Phil Ballard, Jim Tetrud, and Ian Irwin all deserved credit for their brilliant clinical detective work, and if they didn't hurry, they might be left behind.

Warren Grant Magnussen Clinical Center, National Institutes of Health, Bethesda, Maryland

Three thousand miles away, Dr. Stanley Burns was performing a very detailed clinical examination of David Silvey. "I want you to try and flip your hand back and forth twenty times, please." The almost totally paralyzed figure of David Silvey sat in the chair, looking very uncomfortable. He strained and his hand started moving very slowly. After the second turn it stopped altogether. "Okay, just rest for a while," said Burns. David Silvey, who couldn't talk, just grunted.

On returning to NIH, Burns's first step had been to take both brothers off L-dopa. After five days without L-dopa, David had frozen up. He became bradykinetic (slow-moving) and rigid, and all voluntary movement ceased with the exception of his eyes. Burns found essentially the same set of symptoms that Jim Tetrud had seen a month before in Watsonville. It was remarkable that just twenty years ago, before L-dopa, all Parkinson's disease sufferers ended up looking like the Silvey brothers did now.

Helping David Silvey to his feet, Burns asked him to walk back and forth in the room. Slowly and with great effort, Silvey shuffled forward. After a few steps he began to slow down, then he stopped altogether, frozen midact. Burns walked toward him and put his foot in front of Silvey's, positioned across his line of movement. "Try and step over my foot," said Burns. David Silvey, as if by magic, picked up his right foot, stepped over Burns's shoe, and continued forward. This was a classic sign of Parkinson's disease. Nobody knew why exactly, but the act of stepping over something in their field of vision—a line, a foot, or even a paper towel—can enable a person to escape from the impasse.

Once he had done all the tests on both brothers—including a biochemical analysis of their cerebrospinal fluid—and was fully convinced beyond a shadow of a doubt that David and Bill Silvey had full-blown parkinsonism, Burns wanted to reproduce the symptoms in animals. The California addicts had stumbled on a toxin which, while tragic for them, might just lead Parkinson's disease research out of the dark ages. If MPTP caused parkinsonism in people, then it ought to cause it in animals as well. And that would be a breakthrough. Scientists had long dreamed of finding an animal model for Parkinson's disease, which would enable researchers to systematically study the disease in the laboratory and test new forms of treatment efficiently and quickly. Was the discovery of a brain toxin in a synthetic drug the serendipitous event that would make that dream come true? Burns hoped so.

Department of Neurology, Valley Medical Center, San Jose, October 1982

Ever since Lysia Forno's suggestion about writing an article, Langston had been asking colleagues at Stanford and at VMC for their opinions. Should he send the article to a specialist neurology journal like *Annals of Neurology,* or to a general medical journal like *The New England Journal of Medicine*? He certainly did not want to send it to an obscure specialist journal, where it might meet the same fate as the Barry Kidston paper. On the other hand, if he sent it to a prestigious journal like *The New England Journal of Medicine,* with rigorous peer-review procedures, it might take a year or more to get into print.

Scientists publish their research for reasons that seem somewhat arcane to the general public. Unlike publishing an article in a general-interest magazine like *The New Yorker,* no money is paid to the author. In fact, some journals charge the author for the right to publish their work. The author freely gives his or her paper for two reasons, one of which can be considered altruistic (the advancement of science and knowledge) and the other quite personal (peer recognition and career advancement).

When an article is submitted, there are no guarantees that it will be published. First the article is sent out for review by at least two experts in the field to see if it passes scientific muster. If judged scientifically sound by the reviewers, it must still be considered important enough to be published. Journals that employ this process are referred to as "peer reviewed" or "juried" (since a jury of scientific peers must pass judgment on the article). If a scientist's paper successfully navigates this process (revisions are frequently required before final acceptance), then he or she will get credit for the discovery described in the article.

While very time-consuming, the juried journal system is crucial to the entire scientific enterprise. It provides a way of scrutinizing new discoveries before they are published and exposes those discoveries to further scrutiny after they are published, since scientists can then repeat the experiments described and check results. In fact, a new discovery is unlikely to achieve final acceptance by the scientific community until it is independently replicated by another laboratory. All in all, the journal system provides a strong motive for scientists to do good, honest, and accurate research.

To get a scientific paper published in a major peer-reviewed journal is not easy. The researchers have to carefully organize the material and thoroughly check the data. Mistakes aren't just embarrassing, they can ruin a scientist's career.

As he was primarily a clinician rather than a basic scientist, all but one of the papers that Langston had previously written had been clinical papers. These consisted of reports of unusual or interesting cases he had seen. Somewhat ironically, the only scientifically oriented paper he had ever written—coauthored with Lysia Forno—was on Parkinson's disease. It involved counting the number of Lewy bodies in the hypothalamus, an area of the brain that controls hormones.

Because of his workload, that paper, like all his papers, had taken him nearly two years to write. Langston realized that if he wanted to publish a scientific article on MPTP he would have to act quickly. Within a few months, the story would have moved on. If he failed to publish the work, few people outside of San Jose would know about their contribution. The NIMH, which had already published in the *Journal of Psychiatry Research,* would reassume ownership of the field, and as a friend commented, "They would be no more than a historical speck of dust."

Langston was wandering down the hall of the neurology department on the seventh floor of the medical center, pondering this problem, when his friend and colleague John Hotson, who helped him run the neurology department, came by. John noted the puzzled look on Langston's face, and asked, "Why so deep in thought?"

"John, I'm really in a quandary as to where to publish these frozen addicts. It needs to get out fast, but the implications are very important, so I want to put it in a top journal, but they can take forever to get papers out."

John thought for a moment: "Why not publish it in *Science*?"

Never in a hundred years would Langston have thought of this. Considered one of the most prestigious journals in the world, unlike most scientific journals *Science* takes articles from *all* areas of science. It publishes papers which the editors think will have an impact that reverberates across disciplines. Most importantly, compared with other specialized journals, *Science* has a quick turnaround time, it is published weekly, and it is very widely read and cited.

Almost immediately, Langston began to have doubts. *Science* prided itself on only publishing stories of major importance, on subjects from nuclear physics to molecular biology. Was this really so important? Moreover, *Science* hardly ever took clinical papers, but favored basic research.

After discussing it with his collaborators, Jim Tetrud, Phil Ballard, and Ian Irwin, they decided to go ahead. They wrote the paper in three months, sent it off, and waited.

Warren Grant Magnussen Clinical Center, National Institutes of Health, Bethesda, Maryland, November 1982

Sandy Markey was a meticulous chemist. He prepared pure samples of the various compounds in Barry Kidston's fateful synthesis so that they could be injected into rats. Working with his colleague Mike Chiueh, they first injected MPPP, and just as before, rats developed "ratatonia." But, also as in his previous experiments, within three hours they returned to normal. The hydroxyphenylpiperidine produced no effect at all. Next, Markey and Chiueh injected a pure preparation of the by-product, MPTP. This was what Langston had asked about. Markey and Chiueh waited and waited. After four hours, the rats seemed to be completely unaffected. They repeated the experiment with a different dose. Again there was no effect. Markey felt deflated. He called Stan Burns to tell him the disappointing results.

Burns was not surprised that the experiments had failed to produce any effect. As a neurologist, Burns thought it possible that a small animal such as a rat could be insensitive to a neurotoxin that affected a human. There is a vast difference in the way that different drugs affect different animals, and this problem is basic to the search for good animal models for human diseases. A dose of morphine that would kill a man would merely put a dog to sleep. MPPP, which gives a human a heroin-like high, will temporarily paralyze a rat. It was perfectly possible, even likely, Burns reasoned, that a substance which gave humans Parkinson's disease might leave a rat totally unaffected. The most promising approach in Burns's view would be to try out MPPP and all its side products on nonhuman primates. Parkinson's disease was, after all, a disease of primates, human primates to be exact, and it was quite possible that only primates could experience all its manifestations.

Burns had finished his tests on his two human patients—the Silvey brothers. He would have liked them to stay longer, but they had not enjoyed their stay at NIH. On August 20, David Silvey discharged himself from NIH and flew back to California. Bill Silvey followed a few days later. Burns was quite satisfied that they had parkinsonism as a result of a chemical lesion in their brains. Now the time had come to experiment on nonhuman primates—monkeys.

Unlike Valley Medical Center, NIH scientists had easy access to experimental monkeys. Burns was keyed up as he prepared to do his first monkey experiment. Like Markey, Burns intended to systematically try out each side product of Barry Kidston's synthesis. He began with MPTP because its chemical structure was closest to the dopamine molecule. Burns guessed that a dose of maybe 1 milligram per kilogram would be toxic to the monkey, and injected it. Within hours, the first monkey rapidly developed a tremor and died. There was no doubt that MPTP was toxic, but if it was the guilty chemical, he had overdone it.

Burns tried again, using much smaller doses given several times a day. After several days, he began seeing classic but mild parkinsonian symptoms. The monkey's eyes closed partially. Burns had observed the very same "eyelidosis" in the Silvey brothers. Burns tried a third animal with slightly increased multiple doses, and this animal developed full-blown parkinsonism. The monkey was hunched over in a classic parkinsonian stoop, completely frozen. For hours it remained in that position, hardly moving at all. When it did move, it was very slow. Burns observed that the monkey was very rigid in the limbs and neck and had a definite tremor. Then Burns gave the monkey a biscuit and it started eating it. Barely had it started chewing when it froze in the middle of the movement. For several minutes the monkey remained like that, until Burns nudged it to get it going again. All the classical clinical signs of Parkinson's disease Burns had seen in the Silveys were there in the monkey. But what about the neurochemistry?

After anesthetizing the third monkey, Burns took a long syringe and carefully drew some cerebral spinal fluid from the spinal cord of the monkey for analysis. Its composition would either confirm or refute his clinical findings. In the brain, dopamine is metabolized into (among other things) homovanillic acid (HVA). Medical scientists use the levels of HVA in the spinal fluid as a rough indicator of how much dopamine there is in the brain. A low value would indicate low levels of dopamine, which in turn might suggest that the dopamine-making cells of the substantia nigra had been killed by MPTP. The level of homovanillic acid in the third monkey had fallen following injection of MPTP, and remained low, as in Parkinson's disease. Burns could hardly contain his excitement.

Next, Burns gave L-dopa to the monkey, and it reversed the symptoms almost entirely. Within hours the monkey was jumping around the cage, alert and interested in the world outside. MPTP had induced parkinsonism in a monkey, and the symptoms it displayed looked just like those of the Silvey brothers. And L-dopa had reversed all the symptoms, just as it had for the Silvey brothers. As with human patients, the effects of the L-dopa wore off after a few hours. To confirm his conclusions, Burns needed to examine the brain cells themselves. The monkey was "sacrificed" (the research term for killed) and the brain sent for histopathological analysis. The slides that came back were remarkable. The monkey's substantia nigra was virtually gone. So completely had it been destroyed that Burns had to look at brain slides of a normal animal just to see where the substantia nigra was usually found.

Burns knew that he done something very important. He had *proved* that MPTP is a selective neurotoxin, which can cross the blood-brain barrier and, avoiding all other brain tissues, destroy the substantia nigra, the main source of the brain's dopamine, thereby causing the classical symptoms of Parkinson's disease. MPTP did this in humans and it did it in monkeys. Burns had found the Holy Grail: an animal model for Parkinson's disease.

7

Starting Over

It was 6 A.M. and still dark outside. Bill Langston and Ian Irwin sat eating doughnuts and drinking coffee in the cafeteria of Stanford University Medical Center. After speaking to Stan Burns, Langston was having some doubts as to whether he would have much of a role to play in the future of this research. It had been his careful clinical work that had exposed this phenomenon. But from now on this was going to be a basic-research story, and he had no basic research experience. The scientific investigation of parkinsonism in animals would, naturally, involve animal experiments, very careful chemistry, expert pathology, and eventually, complex multicenter patient trials if and when new drugs were discovered for the treatment of Parkinson's disease.

But Langston was absorbed by the scientific questions that had been raised by the investigation. Ian Irwin and he had started meeting early in the mornings, before either of their workdays started. Sitting in the Stanford cafeteria, the pair dreamed of setting up their own research group. Irwin would handle the chemistry. Jim Tetrud might be lured away from his Watsonville practice to do clinical research with Bill. Lysia Forno would do the neuropathology. Perhaps Phil Ballard might stay on after his fellowship finished and join the team. The six MPTP cases—George, Juanita, David, Bill, Connie, and Toby—had brought them together. Why shouldn't they stay together?

As a clinician, Langston would continue to care for these patients and track their progress as best he could. As a researcher, he might be able to turn their tragedy into something positive, a medical breakthrough that might help millions of sufferers with Parkinson's disease, and perhaps the addicts themselves. California is a place where dreams come true. A few miles away from where they sat, two teenagers, Steven Jobs and Steve Wozniak, had dreamed of making and selling a computer an ordinary person could use, and their dream had become Apple Computer, at the time the fastest-growing company in history.

Langston saw one small possibility. On the campus of Valley Medical Center was a two-story gray building that looked like a concrete bunker. Called the Institute for Medical Research, this independent research institute had facilities for basic research and experimental animals. Langston's contract of employment with VMC and Stanford didn't provide for research time, but he decided to talk with the institute's scientific director anyway. The director was very helpful. He said he would be happy to provide some temporary space and a few scrub mice to get Langston started.

When Langston injected the mice with MPTP, however, the drug had no effect. He repeated the experiments again and again, but still nothing happened. Stan Burns had claimed there was simply no way to give a mouse or rat MPTP parkinsonism. If Burns was right, the future of basic research into parkinsonism depended on monkeys, and IMR had no monkeys. Monkeys were expensive, much more difficult to keep and handle than mice, and carried with them all the potential for animal rights demonstrations.

The famous California entrepreneurs, like Jobs and Wozniak or Hewlett and Packard, had started with very little. Working from a garage, legend has it, they begged and borrowed to get started. Langston realized he would have to do the same. Twenty miles to the north, the Stanford Research Institute in Menlo Park had a large primate facility. Langston contacted Chuck Rebert of SRI and explained his problem, and Chuck, in true California hospitality, told him to come on over. Together they injected two squirrel monkeys with MPTP. To Langston's excitement, they developed parkinsonism. Burns had been right. The animal model was exquisite. Chuck agreed to give Langston a monkey so that he could begin experiments at the Institute for Medical Research, but Langston suddenly realized that he wouldn't be able to take it back with him, because he didn't have any monkey cages at IMR. Before he could borrow a monkey, he would have to go to a pet store and buy a cage.

It was pathetic, Langston thought to himself: his basic research program could not get started because he didn't have a cage for his one borrowed monkey. How could they possibly compete against the mighty NIH? Langston thought of the rolling Bethesda campus and all of its resources. There was so much expertise in one place—chemists, molecular biologists, geneticists, veterinarians, laboratory staff, computer experts. There were state-of-the-art animal facilities, with expert technicians. NIH scientists didn't have any problem finding articles; they had the National Library of Medicine on the campus, with five million medical books and journals. These scientists didn't feel isolated. Gathered together on the NIH campus were leading authorities on virtually any area of biomedical research.

Langston heard from a colleague that the NASA Ames facility at Moffet Field, some seven miles away, used monkeys for their space research. As an organization committed to scientific research, they just might consider lending him a cage. So Langston called the man in charge of their monkey research program. It turned out that they were using squirrel monkeys to research motion sickness. When Langston asked to borrow a cage, the man said it could certainly be arranged. Then he said something quite unexpected. "Do you need monkeys? You know, it's really amazing you should happen to call today. We just got a large shipment of monkeys last week and we've been giving them physicals to test their suitability for the program. And twenty of them have flunked their astrophysicals. Would they be any use to you?" Langston was dumbstruck. Fate had delivered twenty monkeys into his hands. Later he would realize it was nearly the equivalent of an NIH grant.

It was a start. But that's all it was. In his mind Langston began to put together a research team. He needed a top-flight chemist like Ian Irwin to head up the basic laboratory. Then he would need clinical researchers to follow the MPTP patients. Perhaps Jim Tetrud and Phil Ballard would be interested. Having a brilliant neuropathologist like Lysia Forno a few miles away in Palo Alto was a real bonus. Of course it all depended on Langston's being able to procure funding. And who was going to fund an unknown researcher?

On December 8, 1982, nearly six months after he had first seen George Carillo, Langston's secretary announced that Elenor Butts was on the phone.

"Who is Elenor Butts?" Langston asked.

"She says she is calling from *Science* magazine," his secretary responded. Langston had almost forgotten about the paper.

Butts's tone was businesslike. "Dr. Langston," she began, "I wanted to ask a few questions about the paper you submitted to *Science*. Why did you submit this paper to our journal?"

Langston's heart was pounding. He knew *Science* usually published only basic research papers, and his was a clinical paper. He took a deep breath, sensing that the answer he was about to give might change the direction of his life.

"There were two reasons. First, we thought this paper would have implications for a wide audience in science, as it deals with a number of different scientific areas: chemistry, neuroscience, neurology, neuroanatomy, and the development of an animal model for a human disease. Second, we felt that the paper had major public-health implications. The designer-drug phenomenon could pose a tremendous threat to public health if it spreads. We felt that maximum public exposure was needed to get the message out."

There was a long pause at the other end of the line. Finally, Butts replied, "We agree." With that, she began editing the paper with Langston right over the phone, a process that is usually done by mail and takes months. The paper appeared in the February 25, 1983, issue of *Science*.

Meanwhile, Stan Burns and his colleagues had been racing to complete preparation for their own paper describing the animal model for Parkinson's disease, which they had decided to send to the *Proceedings of the National Academy of Science*. The reviewers were impressed and early in 1983 the editor informed Burns that his article had been accepted for publication.

It was something of a shock when later on the same day he saw Langston's paper published in *Science*.

After reading the article, Burns was extremely upset. Not only had NIH been scooped; Langston and his coauthors had failed to acknowledge the fact that NIMH had sent the slides of Barry Kidston's brain to them for Dr. Forno to review, slides that were discussed in detail in Langston's *Science* paper. Relations between the two teams, strained since Burns's California trip, now became openly hostile.

Burns felt that Langston had been premature to publish at all. After all, what had Langston shown? Merely that there was strong circumstantial evidence to link MPTP and parkinsonism. Burns, on the other hand, had done a beautiful piece of scientific research, *proving* that MPTP was the cause of substantia nigra cell death and developing a new animal model for Parkinson's disease. Burns knew he deserved the credit, but because Langston had published first, now he might not get the recognition he had earned.

8

Fame and Misfortune

March 1983

The article in *Science* attracted tremendous attention. The local media covered the story on television, and Langston gave another press conference, followed by a dozen or more interviews with newspaper reporters. The *San Jose Mercury News* decided to devote a major portion of their Sunday magazine *West* to the story of the addicts. Scientists, too, were quick to see the potential value of MPTP for research and besieged the Aldrich Chemical Company with orders. Within a few days the company sold out and had to restock.

A week after the article appeared, Langston got a telegram from Sweden inviting him to attend the upcoming meeting of the Swedish Neurotoxicology Society, being held that year in Copenhagen, Denmark. The message said that the invitation was being made at the request of Arvid Carlsson, the distinguished Swedish researcher who had pioneered L-dopa research. Carlsson had read the *Science* paper, and wanted to hear more.

Langston was awestruck. During medical school he had read about Carlsson's pioneering work on L-dopa. As a resident training in neurology, he had flown down to a conference in San Diego just to hear Carlsson speak. Now this giant of neuroscience wanted him to travel to Copenhagen, all expenses paid, to give a talk.

Langston was so excited, he left California without his passport. He realized this in Seattle when he tried to make his international connection. Lisa had to courier the passport to Seattle, and Langston left on the next flight, arriving in Copenhagen a day late.

As Langston listened to the presentations in Copenhagen, his excitement mounted. Hearing the scientists speak rekindled some of the scientific dreams he had had as a teenager. Like most clinicians, Langston had a patient-centered view of medical research. His research in neurobehavior dealt with broad, ill-defined conceptual questions: How is human behavior affected by a lesion in the frontal lobe? How is personality changed by temporal-lobe epilepsy? The answers came not from experiments that other scientists could repeat, but from years of experience diagnosing patients and checking what happened to them. Any conclusions one could draw tended to be imprecise and always subject to question, because of the limited number of cases that came along largely by chance, or the inability to precisely localize the lesions.

The scientists at the Copenhagen meeting, by contrast, asked highly specific questions that were empirically testable in planned animal experiments. The effects of the chemical toxins they were discussing could be measured with surgical precision. Exploring how these chemicals worked was an entirely different process than what Langston had been doing. While their long-term goal was similar to his—to advance science and to help patients—their research concerned molecules rather than vague behaviors and ill-defined brain damage in patients.

Langston was exhausted and nervous when he stepped to the podium. As he told the story of the discovery and showed videotapes of the patients, the audience grew quiet. Afterward a number of people came up and congratulated him. But as interested as they were in the clinical story, what these scientists really wanted to know was why MPTP caused parkinsonism. What was it about MPTP, or any substance, that made it toxic?

These scientists knew that at the molecular level, very little separated a toxic chemical from a harmless one. One chemical compound might be completely safe and do no damage to a brain. Remove a single carbon bond and it became deadly. What interested Carlsson and his colleagues about MPTP was its mechanism of action—how it worked in the brain, how it destroyed the cells of the substantia nigra, and how its action might be blocked. Langston and Irwin had discussed such things in the Stanford cafeteria.

By the time the conference ended, Langston had experienced a revelation. He too wanted to answer these questions. On the plane home, he came up with forty-five different experimental questions he wanted to answer about MPTP. The next morning he began discussing them in earnest with Ian Irwin.

Why, for example, did MPTP seem to be most toxic when taken in multiple doses? The six most affected addicts had taken MPTP several times, whereas some addicts who took just one dose were apparently unaffected. Burns had reported that he needed to give his monkeys a series of doses of MPTP to produce the full effect. Full-blown parkinsonism only developed some seven to ten days after the last dose. Why did it take so long to develop? Ian Irwin speculated that, toxicologically speaking, MPTP might be just the beginning of the story. Usually when toxicity comes from a series of doses, it implies that the source compound undergoes a number of metabolic changes in the body before it does its damage to the brain. Irwin had recently read a paper by the University of Florida scientist Nicholas Bordor which, he thought, might explain these strange properties of MPTP.

Bordor's papers had nothing to do with toxicology; quite

the opposite. He was interested in the fate of medicinal drugs once they enter the bloodstream, and in finding ways to more effectively deliver them to target regions such as the brain. One innovative chemical trick Bordor was investigating involved attaching drugs to special chemical-delivery compounds. In particular, he attached them to a class of chemicals called reduced pyridines. A pyridine is a six-membered ring containing five carbons and one nitrogen. In its fully oxidized form it has three double bonds. If it is reduced by a single double bond, it is called a dihydropyridine. This type of compound is lipophilic (lipid-seeking). Since the brain is made up largely of fats, or lipids, lipophilic compounds slip right into the brain (if they are small enough to penetrate the blood-brain barrier). In contrast, water-soluble compounds, which are hydrophilic (water-loving) tend to be lipiphobic, or fat-hating, and thus cannot get into the brain. Bordor's idea was to attach a lipophilic compound to fat-insoluble medications that normally could not get into the brain. Connected to a dihydropyridine, Border reasoned, they would pass easily through the blood-brain barrier.

What was really ingenious about the concept was that once inside, the drug would become trapped in the brain, providing prolonged exposure to the therapeutic agent. This happened because once inside the brain, the compound gets fully oxidized, turning it into a positively charged compound. Positively charged compounds are lipophobic and can't readily move through the brain to get back out.

Bordor's work gave a strategy for getting good drugs into the brain and keeping them there. Irwin wondered whether it also explained why a very bad drug like MPTP had been able to enter the brain and stay there to wreak havoc.

MPTP is a doubly reduced pyridine (lacking two double bonds), and so in principle should zip into the brain even faster than Bordor's delivery compound. Irwin theorized that inside the brain, MPTP would be metabolized to an electrically charged compound called 1-methyl-4-phenylpyridinium, or MPP+ (which, because it was positively charged, could not get out of the brain). Perhaps MPP+ was the actual toxin. This hypothesis could be tested in the laboratory by giving MPTP to a monkey and looking for MPP+ in the brain. Although MPP+ would be tricky to measure because of its chemical nature, with the right equipment this certainly could be done. If Irwin's hypothesis was right, Langston saw all kinds of possibilities. If the transformation of MPTP to MPP+ was a critical metabolic step, then in principle one might try to prevent that chemical transformation from occurring in the brain. Many of the chemical reactions taking place in the body depended on catalysts called enzymes. Without the help of the enzyme, a reaction simply doesn't occur. One way of preventing the body from metabolizing (or chemically transforming) a substance is to inhibit the enzyme. This approach might work with MPTP, once they knew which enzyme was involved.

For all his enthusiasm and ideas, Langston had a lot to learn about scientific research. Basic scientific research is very different from writing up case reports. For example, in the laboratory, experiments can be methodically planned in advance. In the clinic, one is completely dependent on what cases happen to come through at a particular time. In this setting, diseases are like chance experiments of nature.

Recording data is quite a different thing as well. Like most doctors, Langston was accustomed to writing up clinical histories of individual patients. The style is personal and anecdotal, long and detailed histories of each patient laced with careful clinical observations. Basic scientific research is conducted in an altogether different way. The essential aim is to generate data by a method that other scientists can understand, analyze, and, if they wish, repeat. A scientific study of the effects of MPTP on a monkey could not be written up as a case history such as "Today the monkey Sarah demonstrated subtle but definite tremulousness of her right hand, and a slightly flexed posture which I am convinced is parkinsonian in nature." Since the monkey could not report on its symptoms directly, other ways of generating objective data that could be statistically analyzed had to be found. After writing pages and pages of history on each monkey, Langston quickly came to this realization.

At NIH, Langston suspected Burns would be using the latest electronic monitors. These were devices that used highly sensitive electronic equipment placed in a vest that the animal wore, which sent radio frequency signals to a remote monitor that kept track of the animal's movements. This data was "hard" because it didn't depend on qualitative observations, which by their very nature are at least somewhat subjective. But Langston's team didn't have any such equipment at all, let alone highly sophisticated electronic monitoring equipment.

One of the senior scientists at the Institute for Medical Research, Lou DeLanney, had been showing a lot of interest in Langston's nascent research. DeLanney, now in his seventies, had had a distinguished career as a research biologist. Before coming to IMR, he had been chairman of the department of biology at Cornell and had worked at the famous Jackson Labs in Bar Harbor, Maine, where many of the experimental mice strains that scientists use are bred. DeLanney was an expert on salamanders; in particular he had been fascinated by their powers of regeneration. Amputate a salamander's limb and it will grow back. The same is not true for humans, and DeLanney was investigating why. But his work was going slowly, and a collaborator who had recruited him to IMR was leaving.

DeLanney found the MPTP research fascinating, and it was clear that Langston needed all the help he could get. A lifetime in science working with different animals had made DeLanney a born improviser. When he heard about the problem of the monkey monitor, he called his stepson, a Harvard-trained engineer and computer whiz. They went up and down Silicon Valley collecting parts for a movement monitor of their own making. For a central processing unit, they decided on an inexpensive Commodore computer, which was purchased at a local Toys "R" Us. The completed monitor, which worked superbly, was built at a cost of less than sixteen hundred dollars—far less than the thirty to forty thousand dollars it would have cost if purchased from a scientific catalog.

Langston's team still needed cages for all of their monkeys, since NASA had loaned them only makeshift travel cages, and they needed someone to care for the monkeys on a day-to-day basis. Then they discovered Norm Thompson, who worked for IMR taking care of their mice and dogs. Thompson had worked with primates at UCLA years earlier and relished the idea of working with monkeys again. He was able to locate some old cages in storage at UCLA, drove down to Los Angeles over a weekend in his pickup, and delivered thirty usable cages by the following Monday.

With monkeys, cages, staff, and a monkey monitor, they were up and running. Langston knew his next problem was how to arrange for his own transfer so he could be involved full-time. But even as he set about this, the conservative clinician in him was hesitating about this last radical step. True, the research was exciting. But what if the research went nowhere? Did he really want to resign his position at VMC or his professorship at Stanford?

Cliffside Park, New Jersey, March 1983

Dr. Naokata Yokoyama's attention had been caught by an article in the February 25 issue of *Science*. Something about the title looked familiar and he began reading. After a few sentences his heart began to pound. The substance MPTP was well known to him; he had worked with the compound in the 1960s and '70s on the laboratory bench, synthesizing it as a chemical intermediate. What he had not suspected until now was its toxicity to brain cells. Suddenly, a lot of things made sense to him. Was this the reason why, when he was 37 years old, he had started developing strange neurological symptoms? Was this the reason why his tennis game had begun to fall apart, why his service continually went into the net, why he had started feeling stiff? Eventually, he had gone to see a physician and at age 39 was diagnosed as having Parkinson's disease. For several years he had been on Sinemet and recently he had been experiencing dyskinesias. On reading the article he began to wonder, was his Parkinson's disease caused by MPTP? He sat down at his desk and began to write a letter to Dr. William Langston.

Dear Dr. Langston,
I have read with interest your recent article in *Science*. I am an organic chemist, and from 1964 to 1972 I was working with MPTP and its analogs. During that period I noticed lack of dexterity in my right hand, and upon neurological examination Parkinsonian symptoms were not very clear. But eventually the diagnosis was made when I was 39 years old. For the last ten years I have been suffering from P.D.

I am interested to know if you have noticed any difference between symptoms of usual Parkinsonian patients and those of the victims of chemical lesioning by MPTP.

Of course, [what] I hope to hear from you the most is if there are any signs of remission among MPTP patients . . . Has anyone studied the structure-activity relationship of MPTP analogs for their neurotoxicity? I have synthesized over 100 analogs and I wonder whether some of them were even more toxic than MPTP.

Since the diagnosis of my condition was made, I have been on medication all the time (now I am taking Sinemet and amitriptyline) and manage to work full-time.

Do you have any suggestions as to a medication that is particularly effective for MPTP patients? Are there any diagnostic tests that can differentiate MPTP patients from usual P.D. patients?

I hope that you wouldn't mind receiving this letter and I would be grateful for any information that you would give me.

Sincerely,
Naokata Yokoyama

As Langston read Dr. Yokoyama's letter, interest turned to astonishment and then horror. If Dr. Yokoyama was correct and had MPTP-induced parkinsonism, then Langston might have unwittingly precipitated a public health tragedy among the nation's scientific community. In the introduction to the *Science* paper, Langston had noted that MPTP was commercially available through Aldrich Chemical Company. Within days, close to three hundred scientists had called Aldrich to order MPTP for research, exhausting their stock. What have I done? thought Langston. I might have doomed all these people to parkinsonism and a premature death.

He called Ian Irwin up to tell him about Yokoyama's letter. Irwin was, as always, rational about the matter. "I expect the danger is the free-base form of MPTP. It has a low melting point, somewhere about forty-one degrees Celsius. That's not so different from a warm lab. My guess is that vapors from the MPTP might have come off and Dr. Yokoyama inhaled them. Or it could be through the skin if, say, Yokoyama handled it without gloves. If I had to guess, I'd say he must have inhaled it. He used a vacuum pump. It could have had a leak in the seal, which wouldn't be unusual. The vapor could have emerged into the room and he might have breathed it in."

Langston admired Ian's calm, but unfortunately didn't share it. "So what do you think we should do? Should we get a warning letter out?"

"Absolutely, Bill. You need to go see Dr. Yokoyama to confirm he's got parkinsonism, but in the meantime anyone using the free-base form of MPTP should be pretty careful. They should use gloves and a special hood to evacuate the vapor. In the meantime it's much better if people use the hydrochloride salt of MPTP. That should be much safer."

Anxious to warn biomedical scientists and physicians of the dangers, Langston composed a letter cautioning researchers to be very careful while using MPTP and detailing a procedure to convert the free-base form of MPTP to a hydrochloride salt. Reasoning that many physicians did not read *Science,* he decided to send it to *The New England Journal of Medicine,* one of the most widely read medical science journals.

Cliffside Park, New Jersey, April 1983

As Langston finished a detailed examination of Dr. Yokoyama, there was no doubt in his mind: Dr. Yokoyama had classic Parkinson's disease, with a few minor asymmetries between the left and right sides. Finger tapping and opening and closing his fist were more slowed and irregular on the left. And Yokoyama was experiencing some of the extreme side effects of long-term L-dopa therapy, including excessive movements and rapid fluctuations between on and off.

Dr. Yokoyama was very young to have Parkinson's disease, but it was not entirely unknown for people in their thirties to get it. While it was possible his Parkinson's disease had nothing to do with MPTP, the odds were that he was another victim of the compound.

After the examination, the two men settled down to talk and Langston gathered a full history. Yokoyama had first encountered MPTP in the 1960s. He had been involved in a huge job synthesizing over one hundred benzomorphine derivatives as part of a program to develop new synthetic analgesics. In this work, he frequently needed MPTP as a chemical building block, so he synthesized several kilograms of it, taking no special precautions. He even remembered handling it (without gloves) on the lab bench. Two years after he stopped using MPTP, Dr. Yokoyama got the first signs that all was not well. He went to see a doctor, and got the diagnosis of Parkinson's disease.

Hostelbro, Denmark, Fall 1983

A few hundred miles from where Langston had attended the neurotoxicology conference with Arvid Carlsson, a young

Danish woman read an American magazine account of the discovery of MPTP. She was struck by the symptoms of the California patients; they sounded exactly like those which had afflicted her ex-husband some nine years earlier. Her then-husband, Henning Lunde Nielsen, wasn't a drug addict; far from it. He was a pharmaceutical chemist. She racked her brains to remember that terrible time. It was shortly before Christmas 1975, just after they had got divorced. Henning was working on a big job at the lab. A week or so after he returned to work, everything seemed to fall apart for him. He froze up, his face became masklike, he was unable to speak. He had been taken to the Kommunehospitalet in Copenhagen, one of the larger hospitals in the city, and was initially brought to the psychiatric ward, where he was given an initial diagnosis of catatonic schizophrenia.

Putting down the magazine, she rang up the Danish pharmaceutical company that employed her ex-husband, Ferrosan A/S, to ask what he had been working with the month before Christmas 1975.

The company confirmed that Henning Lunde Nielsen had in fact been working with MPTP as part of a research program to find a new antidepressant drug for Ferrosan A/S. For two weeks he had worked with the compound MPTP, drying it from an alcohol solution. They had no idea that MPTP might be toxic. Over a few days, Nielsen became completely immobile and unable to talk, blink spontaneously, or do anything. After admission to the psychiatric ward, he was seen by a senior psychiatrist, Dr. Welner, who did not agree with the preliminary diagnosis of catatonic schizophrenia. While the patient could not speak or move, Welner soon discovered he could communicate with Nielsen via wrinkling of the eyebrows and an occasional blink of the eye. Nielsen was transferred to the neurological ward and seen by neurologists Erik Skinhøj and Henning Pakkenberg. They realized that the condition resembled parkinsonism and put him on antiparkinsonian drugs, including L-dopa, and he came back to life.

Intrigued, Drs. Skinhøj and Pakkenberg started looking into what might have caused his parkinsonism, and ran a large number of toxicology screens—mercury, lead, carbon dioxide, etc.—but found no clue. They ruled out meningitis via a normal lumbar puncture. They then went to his lab at Ferrosan A/S and looked into what he had been working with. They found several compounds, including MPTP, but since MPTP was not known be toxic at the time, they did not consider it as a possible neurotoxin. Nielsen had remained on antiparkinsonian drugs, the cause of his condition a mystery, until now.

Alerted by Nielsen's ex-wife to the new research in MPTP, Dr. Pakkenberg contacted Dr. Irwin Kopin at NIH. By coincidence, Kopin was coming to Europe to receive an award for his research into the chemical basis of depression. Dr. Kopin, who had seen the Silvey brothers during their stay at NIH, made a side trip to Denmark and he and Dr. Pakkenberg went to Jutland to see Henning Lunde Nielsen in his home. It didn't take Kopin long to decide that here was another case of chemically induced parkinsonism, and he arranged for Nielsen to come to NIH for a full study.

Warren Grant Magnussen Clinical Center, National Institutes of Health, Bethesda, Maryland, Spring 1984

Dr. Stan Burns ran Henning Nielsen through the full gamut of tests, on and off medication. These were the same tests he had used on the Silvey brothers. To test finger dexterity, he asked Nielsen to tap his index finger and thumb together as fast as he could. Off medication, Nielsen could hardly make the movements at all. Then Burns asked him to open and close his hand rapidly twenty times. Like David and Bill Silvey, Nielsen struggled to do so, but the movements were very slow and he seemed to run out of steam. By the sixth go he was unable to make a fist and stopped. Next Burns tested him for rigidity and found him very stiff, not only in the limbs but also in the neck. Like the California MPTP cases, his face lacked any expression.

For some reason, Nielsen's symptoms were worse when he was lying down than when he was sitting or standing up. Burns asked him to try to sit up and get out of bed on his own, but as he tried, Nielsen's limbs began to shake violently. The more he tried to push himself up into a sitting position, the worse the tremor became.

Burns helped him to his feet to observe his gait. Nielsen was bent forward at the hips and maintained this stoop as he "walked." Like the California cases, he moved in small shuffling steps, and his arms did not swing as he walked. Nielsen found turning around very difficult and he could be easily pulled off balance. When Burns stood behind him and pulled him gently backward, Nielsen was barely able to stay on his feet. A harder pull and he would have simply fallen over.

Like the California cases, Nielsen's speech was affected. He could verbalize, but the volume was very low and the speech was very distorted, to the point that Burns often couldn't understand what he was saying. Nielsen's face was masked and he stared with a characteristic reptilian stare.

When Burns had finished all the tests, he put Nielsen back on L-dopa. After a week, Nielsen was a different person. He could get out of bed on his own and move freely. He walked with normal steps and swung his arms as he walked. His stoop disappeared. His reflexes when being pulled backward were almost normal. Most of his stiffness disappeared and he could tap his fingers and clench his fist rapidly. While his face was still a little masklike, Burns was able to make him smile. Now when Nielsen talked, Burns could understand him. There was no doubt: Nielsen was indistinguishable from the California cases. He had MPTP parkinsonism.

San Jose, Spring 1984

As Langston worried about having given a generation of medical researchers premature Parkinson's disease, pressure was mounting on him to decide his future. He knew he was at a turning point. He wanted badly to move to research full-time, but there was no guarantee that he would succeed. Until he had more grants secured, he wanted to hedge his bets and keep his positions at Valley Medical Center and Stanford Medical School.

Stanford Medical School maintained a number of teaching positions at VMC, to train its students and residents. Langston held one such post, which required him to do a certain amount of teaching each week. Unfortunately, it was temporary. Unlike those with tenure at the university medical center, Langston had to reapply for his academic appointment every few years. Usually this was little more than a bureaucratic inconvenience, the outcome a formality. But by spring 1984, he had still not heard about his appointment and began to worry.

The first Langston knew that all was not well was a phone call from Dr. Dominic Purpura, dean of the medical school. Purpura got straight to the point. "Bill, you better come and see me as soon as possible. Your appointment is in trouble."

Langston had no idea what might be wrong. Langston loved Dominic Purpura, whom everyone called Dom. He was a very colorful character, who said *whatever* he thought to *whomever* he wanted, *whenever* he chose. His asides were the stuff of legend. When his chairman of obstetrics had sacked a resident because she got pregnant, Purpura had likened the chairman to Attila the Hun, a phrase quickly picked up in the *San Francisco Chronicle*. While his candor was refreshing to many, it did not sit well with some of the medical center old guard. Langston particularly delighted in Purpura's memos. At the top of his memo paper were two boxes marked "left hemisphere" and "right hemisphere." Depending on the type of memo (whether it dealt with rational, analytic matters or emotional and aesthetic ones), one or the other box was ticked.

After he was seated in Purpura's office, he asked what was going on.

"Bill, the first appointments committee didn't go at all

well, I'm afraid. Somewhere along the line you seem to have made some real enemies, mainly in the department of psychiatry. What in the world did you do to offend them? Also, your evaluations by the neurology residents are really mixed. While some of them loved your teaching and the experience at Valley, others hated it, feeling that you were much too tough on them."

Langston was ready for this one. He knew some of the residents resented being driven so hard at VMC. But he was bound and determined to make their Valley rotation the premier teaching experience, particularly since it was where the best patients were. On the other hand, he was also aware that, as a county hospital, Valley was considered to be a third-rate institution, ranking well below the Stanford University Medical Center and the Veterans Administration hospital in Palo Alto. Langston was determined to change all that, by brute force if necessary. He knew it hadn't been a happy experience for some of them, yet he simply couldn't take their lackadaisical attitudes. Maybe he had been tough on them, but only because they needed it.

Langston explained. "I'm only trying to bash a little clinical knowledge into them. You know what Stanford residents are like these days. They are academic whiz kids, but their clinical skills and attitude can be abysmal."

"Yes, Bill, of course," Purpura replied. "I couldn't make any sense of it. But your problems don't end there." Purpura looked straight at Langston and swallowed. "I don't know if you realize it, Bill, but since your discovery of MPTP you seem to have made a few enemies. There are people at Stanford and VMC that seemingly want to kill your appointment."

Langston had always thought of himself as an easygoing fellow and couldn't imagine that he had enemies, let alone enemies who would threaten his livelihood. On the other hand, Langston had to admit that recently he had fallen out with a number of close associates and friends. Barbara Arons, the head of psychiatry at VMC, was a case in point. Arons was a senior psychiatrist, and during his early days at VMC he had enjoyed a very cordial and even warm relationship with her. Not especially academically minded, Arons had excelled at administration. She had an intuitively brilliant mind, with exceptional political instincts, and a remarkable ability to win battles with the county. It was often joked that she "had the greatest right cerebral hemisphere in captivity." It was Arons that Langston went to when he first conceived of the neurobehavior unit, and she had even obtained funds for the fellowship that Phil Ballard received.

All that had changed after he first heard from *Science* that the paper would be published.

Arons had not been directly involved in any of the work which led to the MPTP paper; in fact she hadn't seemed especially interested in it. But Langston nevertheless felt grateful to her, because he might never have seen George Carillo if she hadn't supported the setting up of the neurobehavior unit in the first place. A week before the *Science* article was to come out, Langston had called her at home in the evening to tell her about the forthcoming paper, explain its potential importance, and thank her for her support. It was a move he would soon regret.

About an hour later, Arons had called Langston back. While she had been quite gracious when he had first called, her tone was now quite different—edgy, harder. Langston was floored at what came next.

"Bill, I have decided that I want my name on the *Science* paper."

At first Langston thought she was asking for an acknowledgment, but it soon became clear that she wanted to be listed as an author. She argued that since she had helped create the neurobehavior unit where George was referred, she deserved credit. When Langston protested that she hadn't done any of the research or participated in the writing of the paper, Arons had become irate, claiming that he had exploited her.

Langston was deeply disturbed, enough to call his department chairman at Stanford the next day, who agreed that it was simply wrong to add her name to the paper. She didn't meet any of the criteria for authorship, and adding her name would in a sense be dishonest. Langston decided to hold his ground. But over the ensuing weeks it became clear that their relationship had changed forever. When the *Science* paper came out without her name, she cut off all contact with Langston.

Langston knew that he had also inadvertently upset his friend and colleague in the neurology department, John Hotson, who had been best man at his wedding in 1979.

Langston had been trying to combine his former clinical work at VMC with the new IMR research, but realized that he was not doing either well. When he received a memo from John Hotson asking him to drop by his office to talk about the future of the VMC Neurology Department, Langston suspected that things were coming to a head. It was a bad sign, Langston thought, that his friend John was now sending him memos.

As Langston recalls it, Hotson got straight to the point. "Look, Bill, I need to know something: are you running the neurology department or not?"

Langston chose his words carefully. "John, as you know, a lot of things have happened, not all of which I'm happy about either. When I recruited you to VMC, I never imagined that anything like this MPTP business would come up. But it has changed the whole direction of my career. I want to do research, I want to spend the next ten years trying to find the cause of Parkinson's disease."

Hotson looked uncomfortable. "Is that a yes or a no?"

Langston paused. "Look, John, I do want to step down from my position as chairman of the department. But I want to try it on a temporary basis. I'd like to keep my office here in the department and have a safety net to come back to in three years if I can't get any funding for the MPTP research."

Hotson exploded. "Just what do you think this hospital is, Bill—A place you can come back to whenever you like? I've been working my tail off while you've been giving interviews to the press and going to international conferences. If you want to spend your time on this kind of thing, that's fine with me. But give me a break. Just tell me if you want in or out."

Langston was stung by the passion in Hotson's voice. Somehow he had lost a friend, probably for good. Hotson had initially been very glad for his success, and had even encouraged Langston to get involved in basic research. Now Langston suspected that Hotson was upset over the time Langston was spending talking to reporters and being a scientist over at IMR. And Langston couldn't really blame him.

If Langston had inadvertently made enemies of Barbara Arons and John Hotson, who else on the faculty of VMC or Stanford might he have alienated? Suddenly, Purpura's statement sounded plausible. Langston looked at Purpura and asked uncertainly, "How serious is it, Dom, do you think I need to start looking for another job"? Purpura laughed. "No, no, of course not, Bill. As long as I'm sitting in this seat you don't have to worry. We will push your appointment through somehow."

Despite Purpura's assurance, Langston felt like his world was falling apart. While intellectually he wanted to switch to full-time basic research, he had become used to his Stanford affiliation and he liked treating patients. Basic research held so many uncertainties. What if he had inadvertently given a generation of medical researchers premature Parkinson's disease? What if his nascent laboratory got blown away by the NIH?

A month later, Langston was in Florence attending an international neuropharmacology meeting at which he was presenting the results of some of their early work on MPTP and its effects in experimental animals. His wife, Lisa, had come with him and they were hoping to enjoy a short holiday afterward, away from medical politics, a break when he could reflect on his future. They were staying in a pleasant little *pensione* and one morning they happened to sit down for breakfast at the same table with another American couple who were attending the same scientific meeting. The husband was a professor on the East Coast.

When Langston mentioned that he held a Stanford appointment, the professor's face lit up. He had some exciting news. "I hear your dean has been sacked."

Langston's heart skipped a beat. "Are you sure it was the dean and not one of the department heads?" "Oh no, I'm sure it was the dean. I'm surprised you haven't heard."

In a major political upheaval, Dean Purpura, his major ally at Stanford, the man who had assured him that his teaching position was safe, had been forced to resign. Without Purpura to defend him, he was doomed. With no powerful friends to stand for him at Stanford, his academic appointment was probably history. And John Hotson's increasing opposition looked likely to end his career as head of neurology at VMC as well. Ironically, after dithering for months about what to do, fate had decided Langston's future for him. There was no way back. He had to risk all on MPTP research.

9

Side Effects

San Jose Jail, Summer 1984

At the moment that Langston decided to leave hospital medicine, George Carillo was suffering badly. Sitting alone in his cell, he saw and felt hundreds of spiders crawling all over his body. Madly, he tried to pick them off. He screamed out as best he could. A guard put his face to the cell viewing window and asked George what was going on. But George didn't hear what he said; he just screamed louder. As far as George could tell, the guard had spiders running all over his face as well.

The guard shook his head and walked on. That crazy George was having hallucinations again. He needed to be seen by Dr. Langston or Dr. Ballard.

For a while, things had gone much better for George. The L-dopa therapy begun in late July 1982 had reversed all of his symptoms, and he could walk and talk almost normally—so much so that the hospital had returned him to the jail to serve out his sentence. Six weeks later, he was back at VMC with severe multiple stab wounds. He had attacked another prisoner and he had got the worst of the exchange. During the first year, Langston and Ballard would see George every few months for one reason or another. Sometimes he would run out of L-dopa, other times he would not take it correctly. In the spring of 1983, George started having hallucinations.

The hallucinations occurred throughout the day and night. They were vivid and terrifying. In April 1983, George began seeing bugs writhing under his skin. He started to pick at his hand manically, trying to take them off. When he was brought into the clinic he saw the bugs on Phil Ballard's face. To Ballard and Langston there was no doubt what was happening: George was experiencing L-dopa side effects. As the summer of 1983 wore on, the hallucinations got worse. Bugs, spiders, worms, and finally snakes terrorized his waking thoughts. Even if the medication was stopped, the hallucinations continued for weeks, although within hours he froze up like a statue.

What George was experiencing was well known to neurologists treating Parkinson's disease, although with ordinary Parkinson's patients, it generally took several years before the side effects developed. Neuroscientists had speculative theories about what was happening at the cellular level, but no firm evidence.

When a patient takes a tablet of L-dopa, he is introducing a substance, levodopa, that will be turned into dopamine inside his brain. L-dopa is given because it crosses the blood-brain barrier, whereas dopamine does not. As currently there is no way to deliver the contents of a pill to just one small area of the brain, this dopamine floods the brain, going everywhere. While much of the dopamine is taken up by the nerve terminals from cells projecting from the substantia nigra to the striatum, thus producing a dramatic reduction in the patient's symptoms, there are nerve cells in other areas of the brain that are sensitive to L-dopa as well, notably in an area called the limbic system. These areas are also rich in dopamine receptors and absorb the large amounts of circulating dopamine, producing various side effects.

The limbic system is involved with emotion. George's vivid and terrifying hallucinations probably resulted from overstimulation of the dopaminergic neurons in the limbic system. A widely held theory of schizophrenia holds that it results from an excess of dopamine, and indeed, George's hallucinations were not unlike the hallucinations experienced by patients with schizophrenia.

When he wasn't hallucinating, George sometimes thought of Juanita and what had become of her.

Rocky Boy Indian Reservation, Montana, Summer 1984

Juanita did not think much about George. In fact, if she thought about him at all, it was in anger. She blamed George for the disaster. He was so careless. It was just like him to buy some bad heroin. But mostly, Juanita didn't think of George. Mostly she just sat smoking cigarettes in front of the television at her parents' house, watching soap operas. She was bored and increasingly depressed. Since the tragedy in 1982, her life had lost all direction and purpose.

Shortly after Langston treated her with L-dopa at VMC, she had been arrested for drug possession and incarcerated for a year at the Santa Clara County Women's Detention Center. After she got out, she moved to be with her family at an American Indian reservation in Rocky Boy, Montana—Juanita's mother was a full-blooded Cree-Chippewa—where her mom, her sister, and her nephews and nieces took care of her. Juanita sat, watched TV, and felt sorry for herself. She had good reason to be depressed. She was plagued with L-dopa side effects.

Like George, Juanita couldn't live without L-dopa. Her body simply froze up. Her limbs became painfully stiff. Yet the L-dopa exacted a price for releasing her from her frozen state. Usually about fifteen minutes after she swallowed the tablets, she could tell the medication was about to kick in.

She knew the feeling so well. It was as if she were a robot and somebody had thrown the switch. The muscles in her limbs and trunk started to move by themselves, at first in small movements, then in larger, violent jerks. Juanita tried to control her body, but couldn't. She started to rock back and forth in her chair in a rhythmic dance, trying desperately to brace herself with her legs. Thirty minutes after she took the medication, the rocking motion approached its peak and she had to hold on to the sides of her chair with all her strength to avoid throwing herself on the floor. Sometimes even that wasn't enough. Sometimes the rocking was so violent that she toppled the chair over and her mother or sister would have to pick her up.

If she was standing up, it took all of her efforts to avoid falling over. To Juanita, it seemed that the L-dopa had somehow screwed up her brain, causing her legs and arms to flail around. To stay on her feet she would step from foot to foot in a bizarre dance step, her knees rising almost to her chin. Because she walked with great big exaggerated steps, the children in Rocky Boy made fun of her, calling her the "horse lady."

So-called dyskinesias are a common side effect of L-dopa in advanced Parkinson's disease patients. As with the hallucinations, they probably result from the indiscriminate way in which L-dopa floods the brain with dopamine. The dopaminergic system of the brain is a highly complex and fragile system. Lining the membrane of the dopamine neurons are sensitive receptors designed to recognize and take up dopamine at the synapse. In normal people, dopamine is taken up at a controlled steady rate. A patient like Juanita, however, went for hours with very low levels of dopamine circulating in her brain, followed by periods (after taking L-dopa) when her brain was loaded with dopamine.

Neurologists like Langston believed that the wild swings in dopamine content were very unhealthy for the dopamine receptors in the motor striatum, eventually making receptors hypersensitive to dopamine, producing exaggerated responses. When Juanita took MPTP, it killed most of the remaining dopamine-making neurons in her substantia nigra. Several years of taking L-dopa had severely affected the dopamine receptors in her brain, making her movements uncontrollable.

Greenfield, California, Summer 1984

Connie sat in front of the television, unable to move or talk. Her mother, Nellie Sainz, came and wiped a bit of saliva from her mouth.

After her L-dopa treatment in 1982, Connie had returned home to Greenfield, where her mother looked after her. Connie had a one-year-old son, Jason, whom she wanted to be with, and thanks to the medication she had a new chance at life. She felt very grateful to Dr. Langston for what he had done and sent him a small framed picture of the Virgin Mary with a touching inscription on the back.

But the L-dopa hadn't given her a second chance. For Connie, it turned out to be worse than the disease itself. Within weeks of taking L-dopa, she developed dyskinesias. Within a year, the medication also began making Connie delirious. She imagined people coming at her with knives, trying to kill her and her baby. Alternately moaning and howling, she would stare at the ceiling, her face contorted in constant fear. Sometimes she would pick up a kitchen knife and run into the back garden screaming, thinking she was chasing an intruder. Otherwise, she would writhe uncontrollably with dyskinesias. Life was becoming intolerable for both Connie and her family. And yet without huge doses of L-dopa, Connie turned to stone.

San Jose, California, Summer 1984

While Langston had closely followed George, Juanita, and Connie, it had been harder to keep track of the remaining three MPTP index cases—the Silvey brothers and Toby Govea.

Since being discharged from NIH in September 1982, both

Silvey brothers had been plagued with L-dopa side effects. In David's case, this was partly his own fault, as he had been taking more than the prescribed amount of L-dopa. His first side effect was dyskinesia, which spread to every part of his body. His arms, trunk, face, and mouth would move involuntarily. The movements burned up so much energy that he started to lose weight. After returning from NIH, Bill Silvey had felt well enough to return to work for a while, and he took employment loading trucks with agricultural produce. But after a few months he began suffering from side effects, not only dyskinesias but also severe hallucinations, and he gave up working.

Both brothers suffered freezing episodes, when the medication suddenly switched off leaving them frozen midaction, unable to move. This was not just embarrassing; on one occasion it had landed them in jail.

The story Langston heard was that David and Bill had been robbing a house and were escaping through the garden with the valuables they had stolen. As they mounted the garden fence, one of the brothers started to feel strange. His limbs began to stiffen, and he started to move in slow motion. His medication was turning off. Halfway over the fence it switched off completely, leaving him well and truly stuck. He couldn't move, he couldn't talk; he was trapped at the scene of the crime in possession of stolen goods.

The officers called to investigate a possible burglary had never had such an easy apprehension. When they saw Bill Silvey straddling the fence, they simply picked him up and laid him in the back of the police car, and took him down to the station. A few months later, David and Bill were in Vacaville jail in northern California. Coincidentally, one of the inmates of Vacaville was Toby Govea.

Toby's medication reversed the parkinsonism, but in time produced awful side effects. The L-dopa partially controlled his tremor, but he had begun to suffer from hallucinations and paranoid delusions. Like David and Bill Silvey, Toby Govea had a long criminal record and was well known to the police for robbery and drug dealing. Like David and Bill

Silvey, the L-dopa side effects made it difficult for him to find work, even if he had wanted to.

As far as Langston had been able to discover, it was the tremor that did Toby in. One day, he robbed a bank in Greenfield. Wearing a ski mask to conceal his identity, he made a getaway with a substantial amount of cash. Toby had been able to conceal his face, but unfortunately for him had been unable to conceal his tremor. The gun which he pointed at the terrified customers moved back and forth rhythmically. They were scared, but despite their fear many of them felt sure they had seen this tremor before. By the time the police showed up, several customers had connected the tremor with Toby Govea. When Toby arrived back at his house, the police were there waiting for him.

Connie, Toby, David, Bill, Juanita, and George were caught on a narrow ledge between two unacceptable alternatives. Without medication they were frozen, encaged in their own bodies, unable to speak. With L-dopa, they could move, but at a very high price—fearsome hallucinations, drooling, and terrible involuntary movements. L-dopa was no more a solution for these sufferers than for anyone with Parkinson's disease. The difference was that most advanced Parkinson's disease cases are over 60 and have lived their lives. These patients were young and had perhaps decades of life to come.

In addition to all his professional worries, Langston worried about these six people who had become his most important patients. He had almost nothing to offer them. There were strategies that could be used—cocktails of medications which combined L-dopa with other drugs that could allow for some reduction in the L-dopa dosage. Unfortunately, these drugs often had just as many side effects, sometimes even worse than L-dopa's. There were drug holidays, where patients were taken off L-dopa for a week or so to try to lower their dosage. But these strategies had very limited and short-term success. And patients often described them as worse than a one-week trip to hell.

Langston hoped that the medical insights their tragedy

had provided might eventually help them. Their plight had yielded a new toxin, MPTP, that had opened a new chapter in medical research. That research might one day transform the prospects of Parkinson's disease sufferers, and end up rescuing these young victims from an unbearable condition—assuming they were still alive.

10

The Cause

Despite the detailed description he gave of the disease which today bears his name, James Parkinson had no idea what caused it. Since Parkinson made his meticulous observations in 1817, finding the cause of the disease has become one of the great quests of neurology. Is Parkinson's disease genetic in origin and passed down from generation to generation, like cystic fibrosis or Huntington's disease? Or is it caused by something in the environment—a virus, perhaps, or a chemical?

Getting at the answers to such questions is the subject matter of epidemiology, one of the most challenging scientific disciplines. Few fields in medicine have generated as much controversy. Epidemiology is not just a question of collecting accurate data. Scientists have to think through every stage very carefully—the questions that are asked; the factors that are controlled for; the statistical methods used to process the data; and, most controversially, what conclusions can be drawn from the data. Furthermore, epidemiologic studies are time-consuming and very costly—often taking years, sometimes decades, to complete. As such studies are equally time-consuming and costly to replicate, it can take decades before a flaw in them is discovered.

Parkinson's disease is especially problematic because it is difficult to diagnose. The misdiagnosis of up to a quarter of the brains of deceased Parkinson's disease sufferers in the British brain bank bears this out.

Such factors have meant that not all of the epidemiological studies of Parkinson's disease agree. For example, studies have found that the prevalence of Parkinson's disease in the population is only 65 cases per 100,000 in Sweden but 187 cases per 100,000 in Rochester, Minnesota. Does that mean that Minnesotans are three times more likely to get Parkinson's disease than urban Swedes? Possibly, but possibly not. There are many factors that might contribute to this apparent difference. For example, the two populations may have a different age structure: there may simply be more old people in Minnesota than in Stockholm, and therefore more people with Parkinson's disease.

Also, the physicians may diagnose Parkinson's disease slightly differently, so as to exclude or include different types of parkinsonism. For example, resting tremor is a usual but not inevitable sign of Parkinson's disease. But some studies have required that cases lacking tremor be excluded, thereby possibly excluding some genuine cases of Parkinson's disease. Conversely, other studies have required only tremor to diagnose Parkinson's disease. Patients with the distinct condition of "essential tremor" thus may have been erroneously included, a problem that has accounted for between 10 and 40 percent of false positive diagnoses of Parkinson's disease in community-based studies (studies that include all members of a defined community, such as a city or county). Essential tremor is not Parkinson's disease, but a benign and often familial condition that is common in the elderly. The hands tend to shake when they are being used for such things as writing or drinking from a cup; this differs from Parkinson's disease, where the tremor typically occurs when the hands or limbs are at rest. The cause of essential tremor is not known, and there is no evidence of a degenerative process in the brain in the disorder.

Several studies have investigated whether the incidence of Parkinson's disease is affected by race and gender—whether it is more prevalent among white people, say, than among blacks, or among men than women. Intriguingly, most studies have suggested that the disease is much less common in blacks than in Caucasians, and this observation is consistent with the experience of most clinicians. However the majority of these studies were done in clinical centers, and were therefore biased by referral patterns and accessibility of care. If blacks had a harder time than whites getting to these specialty clinics, there would appear to be fewer of them with Parkinson's disease, whereas in fact there might be just as many blacks with Parkinson's as whites in the general community.

Community-based studies generally have not found this disparity between whites and blacks. One recent community-based study, a door-to-door survey conducted in Copiah County, Mississippi, revealed that the incidence among blacks and whites over age forty was about the same. However, even this study has been criticized, since tremor was required for diagnosis, and not all subjects were examined by specialists in Parkinson's disease.

Studies examining the prevalence of Parkinson's disease among men and women have produced inconclusive results, although a few studies show a slightly higher incidence in men. In China, however, the disease appears to be twice as common in men. Whether this is a real effect or an artifact is still a mystery.

Despite the current limitations of epidemiological studies, most experts agree that the data suggests that Parkinson's disease occurs throughout the world in roughly the same incidence in both men and women in all societies and cultures, and depends most on age. The risk of Parkinson's disease increases with age. Under 40, about 1 person in 1,000 gets Parkinson's disease. Over 70, 1 person in 100 is affected. By the time a person reaches 100 years old his or her likelihood of acquiring Parkinson's disease has risen to 1 in 40.

But age aside, is the incidence of Parkinson's disease worldwide increasing or decreasing? In Olmsted County in Minnesota, where the Mayo Clinic is located, excellent records are available dating back to the 1930s from which one can derive an estimate of incidence of Parkinson's disease. These records suggest that the incidence of the disease has been relatively stable over the last fifty years. But there are serious limitations in trying to extrapolate from Olmsted County to the rest of the world.

What could explain a disease of aging that appears not to have changed significantly over time, yet is so geographically widespread? Is the cause genetic, or is it some agent or agents in the environment—a toxin or a virus?

Theories

The viral theory of Parkinson's disease After the great influenza epidemic of 1918–20, many people developed encephalitis leading to symptoms of Parkinson's disease. Some scientists predicted that once this particular strain of influenza disappeared, Parkinson's disease would die out as the patients aged and died. The patients died, but Parkinson's disease remained. It is now well accepted that these patients, whom Oliver Sacks immortalized in *Awakenings,* had a special kind of the disease—postencephalitic parkinsonism. This specific kind of parkinsonism did die out, but the more common idiopathic Parkinson's disease seen in the elderly continued at a roughly constant rate.

Chemicals in the environment On several occasions, scientists have speculated that specific substances in the environment might be the cause of Parkinson's disease. One candidate is the metal manganese. For at least a century, it has been recognized that manganese miners who inhaled toxic amounts of manganese dust developed psychological disturbances such as memory impairment, anxiety, disorientation, and hallucinations. Physicians sometimes refer to this acute manganese toxicity as locura manganica ("manganese madness"). It is very common in the mining villages of Chile, for example. The first to suggest a possible link with Parkinson's disease was Irving S. Couper in 1837. Since that time many investigators, including Cotzias in 1958, have noted that chronic exposure to manganese dust produces an irreversible syndrome which is strikingly similar to Parkinson's disease. Victims of chronic manganese toxicity suffer from all of the manifestations of parkinsonism, including symptoms of fixed stare, bradykinesia, postural difficulties, rigidity, and tremor.

Manganese is widely distributed in nature and forms about 0.1 percent of the earth's crust. It is very commonly found in iron ores and in many other minerals, and also in coal and crude oil. Manganese is widely used in industry: manganese oxides have been used in glass manufacture since ancient times. While over 90 percent of the manganese extracted is used to make steel, it is also used in the manufacture of batteries, fertilizers, animal feeds, dyes, wood preservatives, ceramics, and gasoline additives (methylcylclopentadienyl-manganese-tricarbonyl [MMT] is extensively used in unleaded fuels). The majority of these industrial uses generate manganese pollution of the environment.

It doesn't end there. Drinking water represents another possible source of manganese intoxication. Gradual weathering and conversion of manganese to soluble salts results in appreciable concentrations of manganese in rivers and seawater. Only one epidemiological study is available linking manganese-contaminated water with disease. In 1941, Kimura and colleagues reported on a small Japanese community where four hundred dry-cell batteries were found buried next to a well used for drinking water. The manganese content of the well was found to be 14 milligrams per liter—some thirty times the level normally found in groundwater. Out of the dozens of people drinking the water, sixteen individuals developed psychological and neurological symptoms typical of chronic manganese toxicity, and three died.

Finally, foods are a potentially large source of manganese. Cereal grains have a high manganese content, as does tea.

Because manganese is so widespread in the environment, some scientists have speculated that it might be *the* cause of ordinary Parkinson's disease. However, it has been ultimately rejected, for two reasons. First, while the clinical symptoms do include many classical parkinsonian features, victims of manganese toxicity also suffer from severe psychological disturbances (memory impairment, disorientation, anxiety, and hallucinations) as well as dystonia (an excessive increase in muscle tone or tightness that results in fixed posture of a body part). Neither of these features is prominent in early untreated Parkinson's disease.

The substantia nigra, the globus pallidus, and the striatum are part of a brain system that is collectively known as the basal ganglia. Typical Parkinson's disease mainly affects the input side of the basal ganglia, causing degeneration of the dopamine-producing cells in the substantia nigra. These are the cells that deliver dopamine where it is needed in the striatum. When dopamine is no longer present to stimulate the receptors on certain cells in the striatum, the signs and symptoms of Parkinson's disease appear. Manganese toxicity, on the other hand, may or may not affect the substantia nigra, but consistently causes major damage to the striatum and the globus pallidus, the output part of the basal ganglia. This produces a similar set of symptoms but with different pathology.

Another possible environmental candidate that scientists have considered is carbon disulfide. In the nineteenth century, this compound was used in the vulcanization of rubber to make condoms, then a cottage industry in France. The hygiene and ventilation in the shops was very poor, and many of the craftsmen developed symptoms of what was later called parkinsonism. Such symptoms were seen again at the turn of the century, in the viscose rayon industry, in which rayon was produced from wood pulp, a process that used large amounts of carbon disulfide. Quarelli observed in 1928 that some 30 percent of the viscose rayon workers had varying degrees of parkinsonism.

While carbon disulfide is not found extensively in the environment—only small amounts are found in the plume ash of volcanoes—it is widespread in manmade materials. Its excellent solvent properties have led to its use in the production of matches, fats, and oils. It is also used in agriculture as a fumigant and fungicide.

So is carbon disulfide a candidate? Again, scientists have agreed that carbon disulfide toxicity fails to reproduce the clinical and neuropathological picture of Parkinson's disease accurately enough. Carbon disulfide is not exclusively a nerve cell toxin. It also is a skin irritant, causes atherosclerotic cardiovascular changes, and has ophthalmological, renal, endocrine, and reproductive effects. Even in regard to the nervous system, chronic carbon disulfide poisoning leads to symptoms somewhat different from those of Parkinson's disease. Like manganese poisoning, it damages the output structures of the basal ganglia (the striatum and globus pallidus) rather than the substantia nigra.

Plants in the environment In the 1950s, one of the most distinguished neuroepidemiologists in the world, Dr. Leonard T. Kurland of the Mayo Clinic, learned that there was a remarkably high incidence of another age-related neurodegenerative disease, known as amyotrophic lateral sclerosis or Lou Gehrig's disease, on the island of Guam. This disease was associated with both parkinsonism and an Alzheimer's disease–like condition. Collectively, Kurland referred to this condition as ALS-Parkinson's-dementia complex, or ALS/PDC.

During the last two decades, it has become apparent that the disease is slowly disappearing from the island, suggesting an environmental factor is its cause. As this change in prevalence coincided with the westernization of the islanders' diet, suspicion fell on various native foods, in particular the lemon-sized seeds of the cycad "fern tree," *Cycas circinalis,* which the local population consumed as an alternative food source during World War II. After years of research, Peter Spencer and his colleagues presented evidence that an excitatory amino acid found in the cycad plant, BMAA, might be the culprit. Spencer showed that BMAA induces some of neuropathological features of ALS/Parkinson's disease complex in primates after chronic exposure. But interesting though this was, it didn't provide an explanation for the millions of cases of Parkinson's disease in people outside Guam who had never seen a cycad nut. Many authorities have since questioned the cycad hypothesis because people in Guam are exposed to far lower doses than those required to produce changes in laboratory animals.

In short, while perhaps some parkinsonian symptoms can be caused by chemicals and plants in the environment, the cause of Parkinson's disease (and its progressive character) remains a mystery. For this reason, physicians draw a distinction between *parkinsonism* (symptoms with a known cause, e.g., manganese or influenza) and *idiopathic* (from unknown causes) *Parkinson's disease.*

A genetic cause For many years, studies have hinted at the possibility that Parkinson's disease is genetic in origin. A disease that is found all over the world, and that seems to be neither increasing nor decreasing in intensity, might well be inherited.

About 1 in every 220 births produces identical twins. Identical twins have identical genes. If you want to test whether Parkinson's disease is caused by nature or nurture, identical twins provide an elegant way to explore the question. If Parkinson's disease is a genetic disease, one would expect to find that if one twin grew up to get the disease, so would the other. One would expect what geneticists call concordance.

One approach to twin studies is to first identify a large

number of twin pairs in which at least one twin has Parkinson's disease. This is exactly what a team of NIH researchers did. Beginning in 1979, they began advertising through newsletters of Parkinson's disease societies for twin pairs (in which at least one twin had Parkinson's disease) and over an eighteen-month period 121 twin pairs were found. After the researchers examined their medical records, they chose a group of 77 pairs who were available for evaluation by the National Institutes of Health. Next, clinical examinations were carried out to verify the diagnosis of Parkinson's disease. Of the 77 pairs, researchers selected 43 monozygotic (identical) and 19 dizygotic (fraternal) pairs of twins.

This cohort of twin pairs formed the largest Parkinson's disease twin study ever conducted. In 1983, after analyzing all the data, researchers announced their conclusions. The results were dramatic: out of 43 identical twin pairs there was only one case where both twins had Parkinson's disease. This was barely different from chance. A year later, a Finnish study of 18 twin pairs came out which found no concordance at all. To most neuroscientists interested in Parkinson's disease, including Langston and his colleagues, the conclusion seemed inescapable: Parkinson's did not seem to be primarily a genetic disease. It had to be caused by something in the environment.

Back to the environment But what could it be? Something found both in Bangladesh and New York, something present in 1817 when James Parkinson wrote his monograph and still present today. A very common substance (or substances), producing an exquisitely precise movement disorder. A toxin that only seemed to affect the brain and killed cells mainly in the substantia nigra. Something, perhaps, like MPTP?

Clearly, most people had never seen or handled MPTP in its pure form. But MPTP was a very simple molecule, a pyridine, so called because of its ring structure. Pyridines are very common; they are found, for example, in coal, coal tar, asphalt, some vitamins, nicotine, and many plants. The more Bill Langston and Ian Irwin thought about MPTP, the more it became clear that this was a very easy compound to make, intentionally or unintentionally. In fact, after a little research Irwin realized that MPTP could spontaneously form out of three very common ingredients: alpha-methylstyrene—similar to the styrene used to make coffee cups (of which at least twenty billion tons are made each year in the United States); formaldehyde (which is used in practically every chemical process); and methylamine, which is produced when any living thing dies and decomposes. Simply bringing these three compounds together, with or without heat, can produce MPTP.

The substance might not even be MPTP, but something similar. Such compounds might be anywhere—in foods, in the soil, in the chemicals used in making everyday things like newspapers. One tantalizing piece of evidence was right under their noses in northern California, and it involved not people but horses.

In California, there is a weed called the yellow star thistle. If a tomato grower in the Sacramento Valley lets his field go to seed after harvest, the next year it will be covered with this fast-growing plant. In the 1940s, veterinarians noticed that if horses grazed extensively on this weed, they developed a catastrophic neurologic syndrome. After a few days, the horses became rigid, developed difficulty swallowing, started to drool, and eventually died. Similar effects were found in horses in Colorado that ingested Russian knapweed. When the brains of these animals were examined pathologically, they showed a remarkably focal degenerative process, affecting the substantia nigra and the globus pallidus. Because of the similarity of this syndrome, both clinically and pathologically, to Parkinson's disease, neurologists had been fascinated with this disorder for years, hoping it would somehow lead to an animal model for the disease. One neurologist had even spent his entire sabbatical in California trying to find the culprit in yellow star thistle that was causing the syndrome. Now Langston and Irwin had joined the ranks of those who were interested.

At the very least, the puzzle of yellow star thistle toxicity was telling scientists that a very common plant contained the chemicals to produce highly specific neurological damage in horses. The environment, natural and man-made, might be full of such chemicals. Perhaps one of these was responsible for Parkinson's disease.

The discovery that MPTP caused parkinsonism rekindled interest in finding the cause of Parkinson's disease. By 1984, researchers from California to Maryland had begun actively looking for substances in the environment, including yellow star thistle. Investigators knew their search would be complicated by many factors. The 1983 and 1984 twin studies may have indicated that Parkinson's disease was not primarily genetic, but that didn't altogether exclude genetic factors. While Parkinson's disease might not be inherited in a simple Mendelian fashion, there might be more complex patterns of inheritance. Certain individuals might be more susceptible to the disease than others.

Complicating the search was the very real possibility that some environmental chemicals might protect individuals against Parkinson's disease rather than cause it. There was some very intriguing data on smoking, for example. At least nine studies indicated that smokers are less likely to get Parkinson's disease than nonsmokers. No one knew why. Other research suggested that measles, diabetes, and dietary factors may protect against Parkinson's disease. All this raised the possibility that whether or not an individual developed Parkinson's disease might depend on a complex equation of genetic susceptibility, causative toxins, and protective factors. An individual might have a genetically determined susceptibility, be exposed to the causative agent, but not get the disease because he or she was also exposed to a protective factor, such as cigarette smoking. To complicate matters still further, there was the issue of age. Parkinson's disease is above all a disease of the aging brain.

Whatever else is going on, aging plays a powerful facilitating role.

Following the publication of his article on MPTP in *Science,* Langston had quickly come to know many of the leading neurologists in the field of movement disorders. One of the most distinguished of these, Dr. Donald Calne, was also located on the West Coast, but much farther north, in Vancouver. Calne had been intrigued with the effects of MPTP and its implications for the cause of Parkinson's disease. After a number of lively discussions at conferences and by phone on the subject, in 1983 Langston and Calne developed a provocative hypothesis about the etiology, or cause, of Parkinson's disease, which they wrote up and sent to the British medical journal *The Lancet,* which frequently published such provocative pieces.

Their hypothesis was based on two factors. First, it is generally accepted that the amount of dopamine in the brain must be depleted by 80 percent or more before the classic signs and symptoms of Parkinson's disease appear. Second, it is known that as people age, striatal dopamine declines at a rate of 5 to 7 percent every decade. The Calne-Langston hypothesis goes like this: Under normal circumstances, this cell death (5 to 7 percent per decade) will not push us over the 80 percent threshold before we die of old age. However, if a significant proportion of the cells were destroyed by some environmental insult (such as an MPTP-like compound) earlier in life—say, around age 30—then natural cell death would do the rest, leading to symptoms of Parkinson's disease in one's sixties or seventies.

With cause and effect separated by thirty years, it would not be surprising that scientists had failed to identify the environmental cause (or causes) of Parkinson's disease.

And the most attractive feature of Calne and Langston's theory was that it was directly testable.

At the time of the original MPTP tragedy, there had been widespread concern that hundreds of addicts might become affected with parkinsonism. In the summer of 1983 Langston had recruited two Stanford medical students, Josh Novic and Jim Brooks, to track down as many other people as possible who had been exposed to the bad heroin. The pair began by interviewing the six index cases—George, Juanita, David, Bill, Toby, and Connie—and developed criteria to help them distinguish between exposure to MPTP and other drugs like heroin and fentanyl. For example, one cardinal sign of MPTP all six spoke about was the burning sensation at the site of injection. Novic and Brooks went on to identify many individuals likely to have been exposed to MPTP.

Langston realized that this larger group of young individuals who had taken MPTP and still had no symptoms of parkinsonism offered a unique way of testing the hypothesis that he and Calne had proposed in *The Lancet*. Here was a group of patients with a high probability of nigral damage caused by a specific environmental cause—intravenous use of MPTP. The degree of damage would vary, but in some cases it might be substantial. If this group of patients were located and followed over time, not only could they be advised of possible health risks, but the environmental hypothesis could be empirically tested. As this group of patients aged, one would, according to the hypothesis, expect to see early signs of parkinsonism developing as they passed the 80 percent threshold of nigral cell damage.

The Centers for Disease Control in Atlanta had been interested in the "MPTP epidemic" from the start, and after the initial workup they stepped in to coordinate a major effort to locate and follow up these patients. By 1985, over 300 potentially exposed individuals were found and interviewed, leading to a well-defined cohort of 147 cases who, all the evidence suggested, had taken the tainted heroin in 1982. While none of the patients had developed definite clinical signs of parkinsonism, one patient had noticed something highly significant: her handwriting was getting smaller. A bookkeeper, this woman noticed that the size of her entries had begun shrinking after she ingested MPTP in 1982. Langston and his colleagues measured the sizes of the numbers in her books over the two-year period and, sure enough, they were getting smaller. Progressive micrographia, as it is called, is a cardinal sign of Parkinson's disease.

One of the many problems doctors have to deal with when confronting Parkinson's disease is that there is no lab test to prove whether or not a living person has the disease, before he or she has developed clinical symptoms. While the spinal fluid from an advanced case of Parkinson's disease may have a depressed amount of homovanillic acid, a breakdown product of dopamine, such a test will not reveal anything in the preclinical phase of the disease. Nor is it possible to do a brain biopsy to confirm that early Parkinson's disease is present. Much as James Parkinson did in the nineteenth century, today's neurologist must use his or her clinical skill to detect the disease as early as possible. What scientists needed was a way to look into the brain and visualize how much damage had been done so that, well in advance of clinical symptoms, treatment might begin. And in the early 1980s a way was devised—the PET scan.

"PET" stands for positron emission tomography. The University of British Columbia Medical School in Vancouver, where Donald Calne worked, ran one of the few centers in the Western Hemisphere that owned such a technology. PET is the most elaborate, delicate, and expensive of all imaging technologies.

The first modern imaging technique that could depict the inside of the human body in three dimensions was the CAT scan ("CAT" stands for computerized axial tomography). The CAT scan uses X rays to take pictures of many different cross sections or slices of the human body (the Greek word *tomos* means cut). These sections are combined by a computer to produce cross-section images of organs and bones. CAT scans, available since the early 1970s, have revolutionized diagnosis. For example, in serious head-injury cases, CAT scans can quickly and noninvasively determine if there is bleeding inside the skull. Following the CAT scan, magnetic resonance imaging (MRI) was introduced, which was superior at imaging the anatomy of soft tissues. In MRI, the patient is placed inside a powerful magnet that aligns the hydrogen atoms in the body tissues. A radio signal is directed to the body part being examined, temporarily disrupting this alignment. When the radio signal stops, the hydrogen atoms return to alignment, but not all at the same time, since different body tissues align at different rates. A computer measures the change in realignment and converts the data into an image.

Positron emission tomography (PET), which was first used to image the dopaminergic system in the human brain in 1980, is different from CAT and MRI in that it detects not only the anatomy of an organ, but also the status of its metabolic function, such as oxygen consumption and blood flow. It can also be used to detect the accumulation of L-dopa as it is taken up by an area in the brain.

First a Parkinson's patient is injected with radioactively labeled L-dopa, which will pass through the blood-brain barrier, and be taken up by the nerve endings of the surviving "dopaminergic neurons" in the striatum. The L-dopa is presumably then converted to dopamine and stored in the dopamine vesicles of these neurons before it is further metabolized or released. It is labeled with a positron-emitting fluoride.

Positrons are subatomic antimatter particles, with the same mass as an electron but an opposite charge. Being antimatter, positrons exist for only a fraction of a second. As with all antimatter particles, the life of a positron is short and explosive. As soon as it meets an electron, its "matter" counterpart with an equal and opposite charge, the two annihilate each other. Matter and antimatter cannot mutually exist in the same space. The masses of the electron and positron are converted into pure energy, following Einstein's famous formula $E=mc^2$, producing two energetic gamma rays that fly off in opposite directions. In a PET scan, the detectors that encircle the patient are configured only to react to such pairs of gamma rays, emitted simultaneously and traveling in opposite directions. In this manner the number of gamma-ray pairs are counted and the locations of the associated positron-electron annihilation mapped.

Over a period of two hours, the detectors register thousands of events and a computer integrates them into an image which shows a picture of how much fluorodopa is being taken up by the different regions of the patient's brain—in particular by the striatum, where the axons from the substantia nigra terminate. Before PET scanning, the only way to assess brain dopamine was to acquire a piece of the actual tissue after death. Now, for the first time, medical science could assess brain dopamine in living humans.

In 1985, six of the asymptomatic cases in the CDC study (individuals who had taken MPTP but who as yet had no symptoms of parkinsonism) were flown over a period of four to five months to Vancouver for PET scans. If substantial numbers of the dopamine-producing cells in their substantia nigra had been killed by MPTP, then the amount of dopamine reaching the striatum would be reduced. The resulting loss of dopaminergic nerve fibers in the striatum atrophy, thereby reducing the capacity of the striatum to take up fluorodopa. The scientific prediction therefore was that the striatum of these preclinical MPTP patients would not take up fluorodopa as well as the striatum of normal controls.

The PET scans confirmed the hypothesis. As predicted, the uptake of fluorodopa in the striatum was decreased compared to normal controls, but was still greater than that seen in Parkinson's disease. These observations showed beyond much doubt that these affected addicts—most of whom were in their twenties and thirties—were well on the road to getting premature parkinsonism. While they didn't as yet have any measurable clinical symptoms, their striatal dopamine was depleted and, with normal aging, they might well in a matter of a few years hit the 80 percent threshold and develop parkinsonism.

Langston wondered whether the designer-drug maker who had caused all this suffering knew what he had done and whether he had any regrets. He wondered whether the drug enforcement authorities knew where he was now and whether they had any chance of catching him.

11

To Catch a Chemist

If MPTP had changed medical research, it had also played a part in changing the law. The plight of the six frozen addicts had dramatized the dangers of designer drugs, fascinating the press and alarming law enforcement agencies. Drug enforcement authorities had feared that there might be an epidemic of disasters like the MPTP cases, not just in California but in other designer-drug centers like Florida and New York as well. Designer drugs were so cheap and potent to make that they feared the number of people abusing drugs would multiply many times. Vigorous attempts to change the law and close the designer-drug loopholes were in progress in state capitals and in Washington.

At first it had seemed like there was no way to catch and convict designer-drug makers, but the Federal Drug Enforcement Agency (DEA), the police, and a number of Washington politicians grappled with the problem.

Convicting people who make and sell narcotics has never

been easy. Since it is almost impossible to catch someone in the act of selling a drug, in the early 1970s legislation was passed that made possession itself a crime. This meant that substances had to be fully described—including the exact position of every atom in the drug molecule—before they could be banned, which led in turn to the designer-drug loophole. Add a methyl group here, a fluoride there, and one ended up with a slightly different and perfectly legal compound.

Law enforcement authorities had been trying to close this loophole and had come up with two strategies. One plan sought to ban a class of chemicals, thereby blocking a fruitful line of designer-drug synthesis like the fentanyl series. But there was great uncertainty that this approach would hold up in the courts. A smart defense lawyer could challenge what was meant by a "class" of chemicals and probably destroy a prosecution's case with clever arguments. Some lawmakers privately admitted that it might even be unconstitutional, and was therefore unlikely to be tested in court. A legitimate chemist working at a pharmaceutical company could probably just as easily get thrown in jail as an illicit chemist, if the law were really enforced.

Another, more practical, way to approach the designer-drug problem was to work out the intermediate chemical compounds a chemist had to make before synthesizing the final drug and ban those. These intermediates are often called "precursors" to the final product. So, for example, in order to make PCP and its derivatives, one would first have to make one or more precursors, such as phenyl-2-propanone. Banning the precursor 1-cyclohexanecarbonitrile (PCC) ensures that a designer-drug maker synthesizing PCP analogs will break the law at some stage before he has made his target compound, even though that target compound might be perfectly legal. But to prosecute designer chemists, the drug enforcement agents have to catch them midway through their synthesis, in possession of these intermediate compounds. It requires careful timing of a raid, and a lot of luck.

All designer-drug operations depend on a large number of commercially available base chemicals. The DEA had set up a computer system which alerted them when anybody ordered a sizable quantity of one of certain foundation chemicals that any designer-drug maker would need. In October 1984, the DEA had been alerted that a very large quantity of one such compound, bromobenzene, had been purchased and scheduled for delivery in Brownsville, Texas. Putting an electronic beeper on the barrel, all they needed to do was sit and wait for the barrel to be picked up from the warehouse in Brownsville where it had been delivered and follow it to its final destination.

Buying bromobenzene was not itself illegal. The DEA was betting that the bromobenzene would be mixed with other chemicals to synthesize a designer drug, and that the chemist would use established routes and first make intermediate controlled precursors, which were illegal. The size of the bromobenzene shipment suggested that this was a large operation. To bust this operation would create a big splash, a welcome victory in the war on drugs. But to succeed they needed a lot of luck. They had to catch the drug makers partway through their synthesis. If they arrived too early—before they had begun making the designer drug—or too late—after the designer drug was completed—the effort would be wasted. The DEA planned the raid on the house in Brownsville where the shipment had been taken, for 2:00 P.M., hoping the occupants of the house would be engrossed in watching the World Series.

The DEA officers and police burst through the door and placed everyone under arrest. The occupants included two brothers, Gustavo and Hector Olivares-Villarreal, a chemist midway through a synthesis, and the house's owner, Vincent Mason.

Downstairs in the basement they found not one but two homemade laboratories. Quickly, agents seized samples of the chemicals. They knew that if they had miscalculated, Vincent and his partners would be free tomorrow. The DEA would have to replace all the glassware and even resupply any chemicals they had taken. To make a case against any of them, forensics had to find some illegal substance in the mixture.

The timing could not have been better. The powder from one laboratory contained traces of a banned intermediate compound necessary for the synthesis of a PCP analog called PCC. They had caught them in the act. Had they arrived a few hours later, the chemical synthesis would have been complete and all they would have found was the new designer PCP, which was quite legal. Vincent and his men would have gone free.

DEA scientists took longer to analyze the powder found in the other laboratory. The target product had been an analog of MPPP called PEPOP. But mixed in was an unidentified contaminant. After conferring with Ian Irwin at the Institute for Medical Research in San Jose, they sent him a sample of the compound. Irwin analyzed it in his mass spectrograph. He had encountered this compound before: it was MPTP.

San Jose, March 1985

Langston was rather surprised when he got a call from the department of corrections in Brownsville, Texas.

"Er, Dr. Langston, one of our inmates has developed a medical problem which he thinks you might be able to help with."

Langston didn't know anybody in Brownsville, Texas, and couldn't imagine what he was talking about.

"What is the name of this prisoner?"

"It's a Mr. Vincent Mason."

There was something vaguely familiar about the name, but Langston couldn't place it. "What has all this to do with me?" he replied.

"Mr. Mason was arrested for manufacturing PCP. He had set up a designer-drug operation down here. And as far as we can tell, he had been active in the Morgan Hill area in 1981 and 1982."

"You don't mean he's the guy who made the bad heroin."

"We think so. We have been in touch with Jim Norris of the Santa Clara County Crime Lab and he told us about your work with the addicts who got Parkinson's disease."

Langston couldn't hold himself back. "What's he like?"

"Well, how would you like to find out?"

Langston didn't understand. "What do you mean?"

"Well, we were thinking of bringing him up to see you next week to give you a chance to examine him."

"Examine him? Why, what's wrong?"

"Well, he's complaining of what might be the early signs of Parkinson's disease."

March 22, 1985

When Langston got up, he knew that he would remember the day for two things. One, it was his forty-second birthday. Two, he was due to meet Vincent Mason. Langston was excited about meeting Mason, but was still very worried about the toxicity of MPTP and wondered how many other individuals might have been affected, not to mention Ian and himself. He now knew of several cases which proved MPTP could be inhaled or absorbed transdermally. There was the Danish chemist Stan Burns had examined, and Dr. Yokoyama. He had also learned of two drug addicts in Vancouver who had snorted MPTP like cocaine. Both had been severely affected with parkinsonism, and one had died.

If Mason had handled MPTP on the lab bench, it would not be surprising if he had contracted it as well.

The prospect of meeting Mason had sent a ripple of jittery excitement through the office. What was he like, this man who allegedly had maimed so many young people with his tainted heroin? Langston had not pictured the individual the police delivered to his office at 11:30. A pleasant-looking man in his late forties or early fifties, Mason was well-dressed, with carefully cut dark hair graying slightly at the temples. He was clean-shaven with an engaging smile, and what turned out to be at times a charming manner.

The two men spent the next three days together. While the ostensible purpose of the visit was a clinical examination, Langston was bursting with questions. He wanted to know everything about this man—where he came from, why he got involved with what he did, what responsibility, if any, he felt. The details of what happened. The two men's fortunes had become intertwined. But for Mason's drug operation, Langston might still be running neurology clinics at VMC.

Vincent Mason told Langston the story of his life. Apparently, he needed to tell it as much as Langston needed to hear it. He knew that he faced five to ten years in jail. He knew that people suspected him of crippling the California addicts. Yet, as he would argue repeatedly to Langston, he wasn't to blame. He invited Langston to check out any part of his story, and gave him full permission to repeat it to anyone.

Vincent Mason was born and grew up in Brownsville, Texas. He had done well at school and, even though his family didn't have a lot of money, he had managed to earn enough to go to law school. Back then, in the 1960s, his future had looked full of promise.

Around this time his parents, whom he was close to, moved to northern California. Vincent, who liked the California climate, decided that he would follow.

"When you moved to California, did you have any thought of getting involved in a criminal enterprise?" Langston asked.

"Not at all," Vincent replied. "I was a lawyer, I loved the law."

"Why then didn't you continue working as a lawyer in California?"

"I wasn't qualified to practice in California and I didn't have the energy to sit the California bar exam, so I looked around for something else to do. I attended some university classes. That was where I met Dr. Galaza."

"Dr. Ernesto Galaza?"

"That's right. He came to lecture. He was a great man,

very inspirational. He had devoted his life to helping the poor and underprivileged. So I started to work with him. I helped him to start the Alviso Clinic."

Langston could not believe what he was hearing. The man who hc suspected had crippled Connie was sitting in his office claiming that he had helped found the Alviso Clinic, one of the oldest and best-known neighborhood health centers for the poor in the Bay area.

"How were you involved?"

"I helped raise the money for it and was the first executive director of the clinic for about nine months."

Langston was dumbstruck. This was beyond belief, but he would soon learn it was absolutely true.

"I liked the work," Mason continued. "After that I was involved in obtaining federal subsidies for low-cost housing from HUD."

"Anything else?" Langston asked skeptically.

"Yes. I worked on a grant for a model cities program with Hank Rosenden."

"Hank Rosenden, the former recreational director of San Jose?"

"Right."

Irony of ironies, thought Langston. He knew Rosenden personally. Rosenden was now executive director of the Peninsula Parkinson's Support Group! It would be easy to check out this part of his story.

"What happened then?"

"I got fed up with the politics of public service and became interested in media—Spanish media to be precise. I bought a share in a radio station."

Langston listened as Mason continued his story. Even though he had great trouble believing any of it, he recorded the conversation and asked questions in a methodical manner. Mason continued, telling Langston that he had operated the station successfully until 1982, when he sold out. It was after this that the events happened which led to the designer-drug tragedy.

Before they could get to that part of the story. Lisa Langston came in with a cake. The people at the lab wanted to celebrate Bill's birthday and had made all the arrangements, not knowing about Mason's visit. As Mason was a guest, Lisa offered him a piece as well. It was bizarre, Langston thought, that they were all standing around sharing cake with the man who was part of a designer-drug operation that had ended up wrecking so many lives.

During the break Langston slipped out to a free phone and made a few calls. First he called Hank Rosenden, who confirmed that Mason had worked with him in 1973 on the model cities program. Then Langston called the Alviso Clinic, which put him in touch with Roy Hernandez, a historian writing a book on the history of the clinic. Hernandez confirmed that Mason did work with the group that founded the clinic under the leadership of Ernesto Galaza, playing a key role writing proposals and organizing the effort. According to Hernandez, Mason was highly regarded among the volunteers setting up the clinic.

Langston was stunned. How could this dedicated public servant, who had given his energies to helping the poor, be part of the designer-drug tragedy? Over the next two days, Vincent Mason told Langston an almost unbelievable story, in which he continued to deny any direct responsibility for the tragedy which had occurred in northern California. But for the confirmation of Mason's work at the Alviso Clinic, Langston would not have believed any of it.

According to Mason, before 1980 he knew almost nothing about drugs. He had friends who used them, but he hadn't been especially interested. Then one day in 1980, when meeting a group of friends for coffee, the conversation had turned to drugs. One of the group casually mentioned that he had just bought a gram of really good cocaine which had cost him $150. Mason remembered being astonished at the concept—$150 for a gram. He had thought it utterly remarkable that some dealer could get so much money for such a small amount of chemical. It was too bad, he thought, that drugs were illegal, because they were obviously a terrific business.

A few days later, Mason had been driving down the Old Monterey Road and had stopped at a drugstore for a Coke. He had been browsing through a magazine rack when he saw an alternative newspaper. Intrigued, he bought a copy and started to read. He found the classified advertisements especially interesting; one in particular caught his eye.

> FOR SALE: Cocaine—$20 per gram, produced under a federal grant, and fully approved by the government.

Mason was fascinated. He knew that there had to be a catch, but something drew him in. He sent off a letter, together with a check for twenty dollars, to a post office box in Los Altos Hills—a wealthy area just south of Palo Alto—and waited. Soon he got a call from an individual who identified himself as Mike Lantle, the author of the advertisement. According to Mason, Mike Lantle was a physicist with an aerospace company, who lived in the fast lane. He liked money, women, fast cars, and skiing at Lake Tahoe. His work wasn't enough to support his fast life. So he had started a scam. It was brilliant. When people who were either too stupid or too greedy to know better sent in their twenty-dollar check, he sent this cryptic note in reply.

> Could you possibly believe that the government would fund the production of cocaine to be sold to the general public? You must really be stupid. Cocaine is illegal to possess. You could be arrested if you had some in your possession.

And that was it. People who sent their money felt like idiots, but they had little recourse. After all, they were sending in money for something that was illegal in the first place. Who were they going to complain to? Lantle had been running his scam for some time. Once he had even got a request from a pharmacist who asked for 10 kilos of government cocaine.

But there was something about Vincent Mason's letter that interested Mike Lantle, so he called him up and invited him to lunch. The pair hit it off immediately and began to meet on a regular basis.

In 1981, when Vincent was selling the radio station, Mike had come to Vincent with a proposition: Would he like to invest some of the money from the sale of the station in a novel new venture? One of Mike's friends in the soda pop business had gone broke and wanted to sell his manufacturing plant. Mike's idea was to make a soft drink with a difference: instead of using carbon dioxide in the pop, he wanted to use laughing gas. Laughing gas, or nitrous oxide, was one of the first recreational drugs. In the nineteenth century, the rich and famous flocked to laughing-gas parties, where they inhaled the gas along with beverages.

Mike was proposing that they manufacture a soft drink that was not only refreshing but intoxicating as well. The best part—the one which appealed to the lawyer in Vincent—was that it would be legal. Laughing gas wasn't a scheduled drug. They would call their new product "Giggle."

According to Vincent, they had bought some chemicals and begun to experiment making Giggle at a house in Saratoga. Mike knew a lot about drugs and the law and was actually planning to write a book about making drugs legally. The concept of legal designer drugs fascinated Vincent. Here, surely, was a way to have one's cake and eat it too: the vast profits of narcotics and the safety of a legitimate enterprise. Mike told him that there were chemists out there making all kinds of stuff and that the law would never catch them.

Then had come the first big setback. Mike's experiments did not go well, and there had been an explosion at the house in Saratoga. The fire department had come and found the huge quantities of chemicals, and even though they hadn't been doing anything illegal, the police had started asking them both questions. Mike not only lost his truck in the fire but shortly afterward also lost his aerospace job. He rang up Vincent asking for some money to cover his losses, and the pair argued. Mike slammed down the phone and disappeared from Vincent's life.

Langston listened calmly. Nothing that Mason said had the ring of truth about it. A narcotic drink called Giggle. It sounded more like a hoax than a confession. But back in the summer of 1982, Jim Norris of the crime lab had mentioned something about a chemical fire in Saratoga.

"So what happened then?" Langston asked.

"After the fire, I moved to Morgan Hill where I rented a house. I got on with my life until I heard from an acquaintance of Mike's. He was looking for lab space for a chemist he knew who had discovered an allergy drug.

"He wanted to start his own lab and keep the profits for himself. He asked me if I would be interested in making some easy money."

"Are you saying that you had no idea that illegal drugs were involved?"

"Absolutely. I wasn't looking to get into anything heavy."

Mason went on to tell Langston about Carl, a brilliant and high-strung chemist. Carl was twitchy and quiet.

A procedure was established whereby Vincent would pick Carl up from Fremont Bay Area Rapid Transit Station and drive him back to his house. Carl would tell Vincent what chemicals and equipment he needed, and Vincent would go down to the San Jose Municipal Library and look them up in the chemical catalogs and order them directly. For this service Vincent would charge a 10 to 20 percent commission. He remembered ordering large quantities of substances like bromobenzene from companies like the Aldrich Chemical Company.

Carl brought copious notes and photocopies of articles and immersed himself in the task at hand. Vincent recalled that Carl would sometimes sit for hours crouched on his haunches, staring at his glassware. Sometimes he would fall asleep in this position. Gradually Vincent started to spend time with Carl, who seemed to be lonely and depressed.

After four failed attempts, Carl successfully made a small batch of the chemical he was working on and took it away to analyze with a mass spectrograph. When he returned the next day, Carl had a printout with him and looked pleased with himself. "It's the real thing," he said. "It worked!"

Mason remembered that Carl had made the first large batch of the drug on May 20, 1982, which happened to be Mother's Day. The crystals emerged from the synthesis still wet and smelling of acetone. Carl said a proper lab would have had a drier to finish the job, so in Vincent's makeshift lab they had to improvise by spreading the crystals out and leaving them there to dry. As Carl couldn't be there all the time, it was Vincent's job to come in every few hours, crush the crystal with his hand, and smell to see if the acetone smell had gone. He may have literally been inhaling MPTP, Langston thought.

According to Mason, by early June they had made and shipped a large amount of material. Things seemed to be going very well, and his associate was pleased.

"When did you first realize that what Carl had made caused people to freeze up?"

"I never made any connection between that and what Carl was doing. As far as I knew, he was making an allergy drug; that's what I'd been told."

"Are you saying you suspected nothing?" Langston asked.

"In retrospect, I remember one or two strange things. One day Carl became very agitated and distraught. He kept shouting on about how he had damaged someone, ruined her life. At the time I didn't understand what Carl was going on about. But I suppose he had just heard about the parkinsonism."

"What happened then?"

"I never saw Carl again. And a bunch of fire marshalls came to inspect the house. Of course they didn't find anything, but they ordered me to get rid of the chemicals."

"You told them, didn't you, that you had been using the chemicals to test new flavors for sno-cones?" Langston said.

Without hesitating, Mason replied, "That was true. I often made sno-cones for children on hot days. In fact, an aunt was visiting me that day with her family, and had been experimenting with some new flavors for the children earlier in the day."

Langston thought to himself, this guy has an answer for everything. But he was also somewhat impressed. Hard to believe as the story was, with its wild improbabilities, try as he might, he could find no internal inconsistencies. Langston probed further, taking Mason over the same ground again and again, hoping to expose inconsistencies.

Like many physicians, Langston was an expert at this. He had years of practice in evaluating case histories of possible malingerers who faked illness. One of the clinical arts in differentiating malingerers or in diagnosing hysterical illness was to find inconsistencies in patients' stories. But there were no inconsistencies in Mason's story. To repeat such a convoluted and intricate story without ever altering it or making an error was quite a feat, and something Langston had not encountered before. But then, Mason was a lawyer.

Mason continued his story. That night, the man who had sent Carl to him originally had sent a truck to remove all the equipment and chemicals—including some 200 to 300 gallons of ether and acetone—and take them to an address in King City.

The next morning—the third day of the interview—Mason came in and claimed to be depressed. He was depressed because he feared that Langston didn't believe him. This put Langston back on the defensive. Langston, anxious to hear the last part of the tale, encouraged him to continue.

A few months after the fire inspection, Mason ran into a man who offered to sell him a new formula for a designer PCP based on an antihistamine. The price was cheap, two hundred dollars. The potential profit was enormous, in the millions of dollars. Mason got the idea of taking this formula to Brownsville in Texas and setting up his own operation there. He had become a little too well known to the

California police for his own good. In Brownsville he should have a much easier time. Now, for the first time, Mason appeared to be admitting some connection to the designer-drug business, after denying that he had known about it when it was going on in his own garage. Langston found this too improbable to swallow, but continued listening with fascination.

Mason continued his story. He had gone back to Brownsville, met Dr. Gustavo Olivares-Villarreal at a party, formed a partnership with Gustavo and his brother Hector, and started to spend money. Gustavo had tried to synthesize the designer PCP but had failed. So eventually Mason had found a Mexican chemist, ordered up large quantities of chemicals—something he knew very well how to do—and set up a lab to make a legal form of PCP.

The lab was set up in a small house he had bought for his parents to retire in when they expressed their intention to return to Brownsville. In fact, his parents decided to stay in California. Mason had spent a lot of time and money on this operation, but he knew that if it worked he would never have to work again.

"What happened then?" asked Langston, on the edge of his seat.

"The chemist had quite a job trying to get the formula to crystallize. He was pretty good, though. He had a lot of tricks."

Langston had heard Ian Irwin talk about the tricks chemists used to make things happen. Of course, chemists had to be meticulous in the way they measured things out, but over and above this meticulousness was a body of tacit knowledge not to be found in any textbook. The chemist had managed to crystallize the end solution, but it hadn't been easy. Every schoolchild can take a pure solution like copper sulfate and get it to form blue crystals. But a mixture of two or three chemicals will not crystallize so easily. The chemist had struggled for three days, trying different strategies to get the mixture to crystallize. He had dipped his stirring rod into the solution, removed it, and let it dry. After a few minutes there were a few light crystals stuck to the surface. Carefully, he broke off one or two and dropped them into the liquid. Often these crystals seed the crystallization, causing the whole mixture to crystallize within seconds. But this hadn't worked. So he had taken a stirring rod and scratched the glass surface of the vessel. The irregularities in the glass seeded the process. A few crystals formed, setting the process on its way. The liquid clouded up as the crystals formed. Within thirty seconds it was all over.

With the test over, the chemist had settled down to make the designer PCP in bulk. The day the DEA raided, Mason and the Olivares-Villarreal brothers had been watching the World Series on television while the chemist worked away in the makeshift lab. The synthesis was nearly finished, but some of the precursors for the designer drug were still present in the batch. Mason and his friends had been nailed.

At the end of three days of story and a full clinical examination, Langston didn't know what to think. Mason had not even mentioned that the DEA had actually found a second laboratory in his Brownsville house that was making meperidine analogs contaminated with MPTP. Langston had doubts about both his story and his parkinsonism.

As for the story, Langston suspected that parts were true and parts invented. At no stage in the interview had Mason shown any remorse for what he had done or accepted any responsibility for the tragedy. Clinically, Mason had no pronounced symptoms, but perhaps he had sustained damage to his brain that would later result in premature Parkinson's disease.

As Langston returned Mason to the custody of the marshalls, the two men shook hands and returned to their separate, now very different lives—Langston to his science and patients, Mason to prison to await trial. Afterward, Langston spent many hours trying to understand how a man with such a background and history of helping others had come to such a state. In the end, Mason decided to turn state's evidence and in return for information received a reduced ten-year sentence and a $20,000 fine for conspiracy to make and possess PCP. Gustavo got fifteen years and Hector thirteen years.

By the time Mason began his prison sentence, MPTP research had yielded its first therapeutic advances.

George Carillo lying "frozen" in the neurobehavior unit of the Santa Clara Valley Medical Center in San Jose, California, in July 1982. At the time, doctors were baffled as to the cause of Carillo's inability to move or speak.

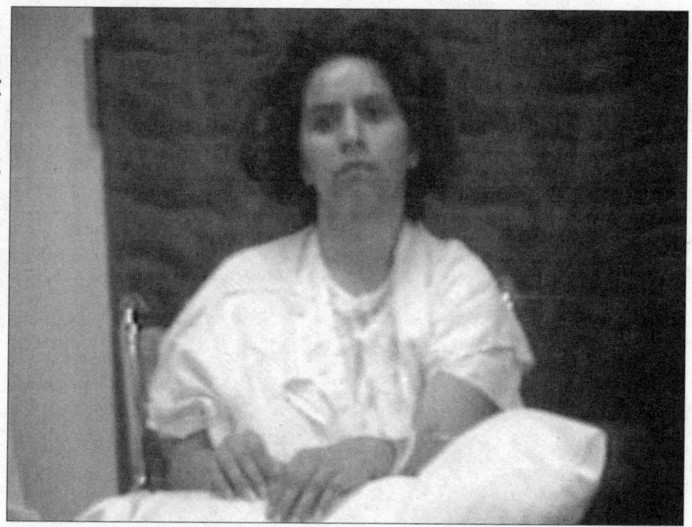

Connie Sainz, the most severely ill of the six MPTP patients, was first taken by her sister to Stanford University Medical Center, where psychiatrists mistakenly diagnosed hysterical paralysis. They sent Connie home, still frozen, saying that in time she would snap out of it.

While the drug L-dopa enabled Juanita Lopez to move, it soon began to produce terrible side effects, among them violent body twisting that she couldn't control (dyskinesias).

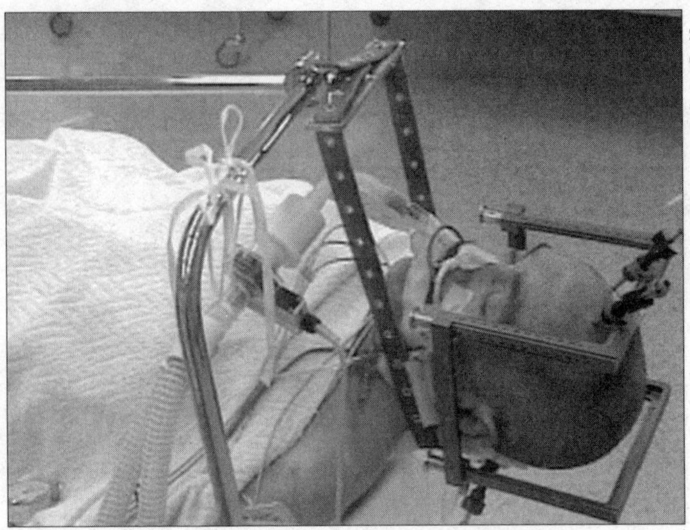

August 22, 1989, 8:00 A.M. George Carillo lies anesthetized in an operating room in Lund, Sweden. The surgical team has mounted a stereotactic frame on his skull in preparation for grafting fetal tissue into his brain.

Courtesy Russ Lee

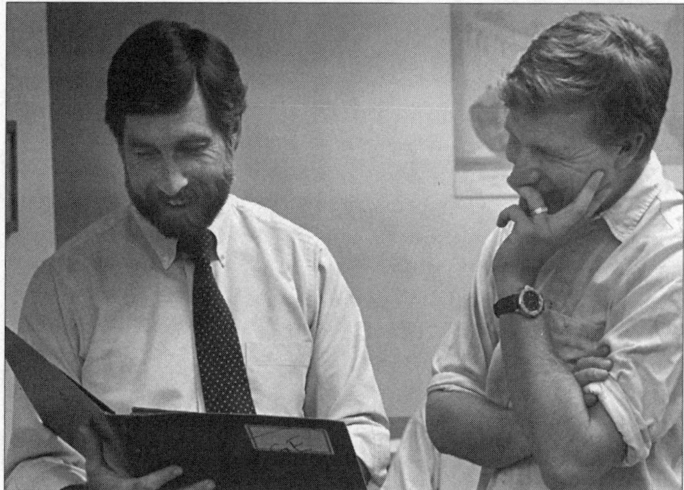

Drs. Bill Langston and Håkan Widner enjoying a photo album of George Carillo's trip to Sweden for fetal-tissue transplant surgery.

Courtesy Russ Lee

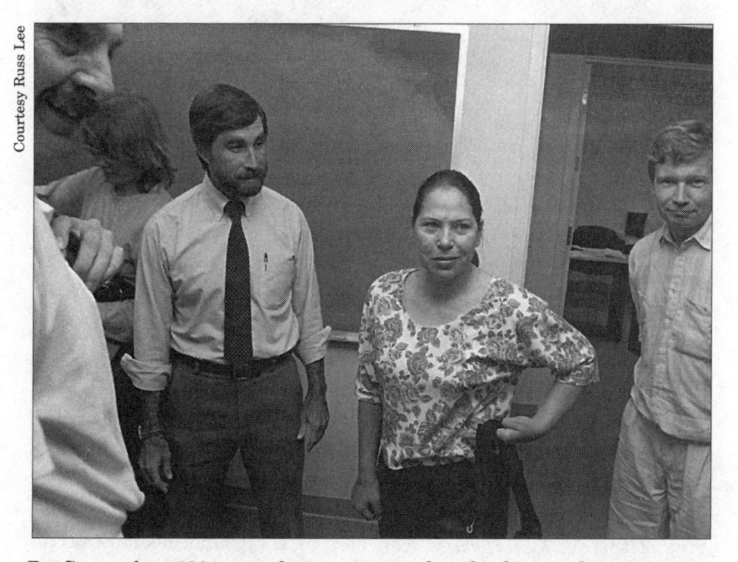

By September 1991, nearly two years after fetal-transplant surgery, Juanita Lopez had dramatically improved, and Drs. Langston and Widner decided to reduce her L-dopa medication by 70 percent.

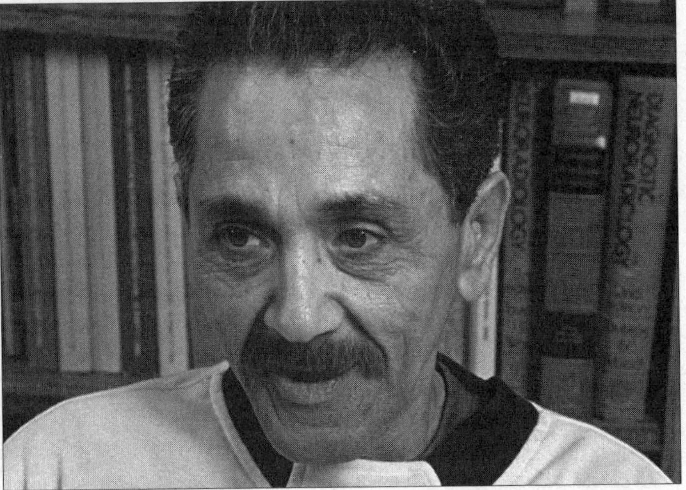

Courtesy Russ Lee

Two years after his surgery, George Carillo had regained many of his motion skills, his facial expressions, and his ability to talk.

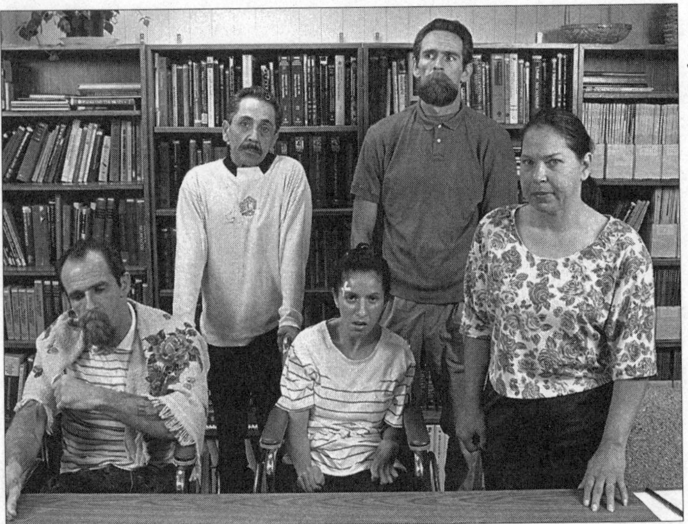

Courtesy Russ Lee

Five of the six MPTP cases seen here at the California Parkinson's Foundation in Fall 1991. George Carillo (standing left) and Juanita Lopez (standing right, front), the two recipients of fetal-tissue transplants, contrast with the still-very-sick patients: Bill Silvey (seated left), Connie Sainz (seated center), and David Silvey (standing right).

12

Trail of Ironies

It was ironic. The designer-drug operation in the Morgan Hill House had started a research revolution. With progress in sight, there would be much more money available for Parkinson's disease research. But there would be a lot of competition for the funds.

Scientific research takes money. Money to pay salaries and benefits, money to buy equipment and animals, money for insurance, money for everything from computers to paper clips. Langston had realized that a major piece of being a basic scientist was raising research funds, and he was spending an increasing amount of his time writing grant applications.

Money for research comes from many sources—federal grants, charities, private foundations. The skill of "grantsmanship" comes in finding the appropriate sources for a given research project. The funding bodies have definite objectives, which may range from a general support of biomedical research (like NIH's extramural funds) to the specific support of a particular disease, like cystic fibrosis. To tap into a particular source of money, a scientist has to convince the funder not only that he or she is a competent scientist, but also that the research furthers the funder's goals.

In 1984, there were four national organizations in the United States dedicated to disseminating information about Parkinson's disease: the Parkinson's Disease Foundation (PDF), the United Parkinson's Foundation (UPF), the American Parkinson's Disease Association (APDA), and the National Parkinson's Foundation (NPF). These bodies also supported some research projects, but were not known for granting large amounts of money. Partly this was tradition, but partly it was because very little promising research had been going on into Parkinson's disease. Judy Rosner, the executive director of the United Parkinson's Foundation, had been making small disbursements for years, but unknown to the scientific community, she had been stashing away money in a special fund. A formidable personality, she had a dream that one day she would come across a piece of research which held the potential to revolutionize Parkinson's disease research—as revolutionary as L-dopa had been in its day. Then she would use her accumulated fund to act.

In common with her colleagues in the other Parkinson's disease foundations, Judy Rosner attended the American Academy of Neurology meeting in San Diego in 1984. As usual, she set up a booth to give out information to interested parties. Langston happened to pass her booth and stopped to chat. The two hit it off immediately. When she heard Langston describe the discovery of MPTP and the possibilities it opened up, Rosner began to wonder whether she had finally found what she had been waiting for. She invited Langston to apply to her for a major research grant.

A few months later, on July 24, 1984, the UPF announced they had decided to fund Langston's group for three years. The size of the grant—nearly half a million dollars over three years—stunned the Parkinson's disease research establishment. Langston had secured enough money to support his full-time research, and his success now made him the target of envious sniping from the old guard of the Parkinson's disease research establishment. A number of eminent neurologists who had spent their lives working with Parkinson's disease didn't see why this upstart Langston should get so much money so soon. It really didn't seem fair.

As Langston read the grant letter from Rosner, he thought about the many things that had happened since he had first seen George. Back on that hot July afternoon in 1982, he could never have imagined that the frail figure of George Carillo would lead to so many changes: changes in his own life and career; changes in Parkinson's disease research itself. That case, and the ones which followed, had opened a new chapter in neurology and given Parkinson's disease research a much-needed shot in the arm. Now that there was an animal model for studying the disease, large numbers of basic scientists—chemists, molecular biologists, epidemiologists—were bringing their powerful skills to bear on this mysterious and tragic disease. Now entire conferences were being organized around a toxin produced by chance in a back-street designer-drug laboratory, MPTP.

Langston thought a lot about MPTP. It had wrecked lives, it had changed lives. It had taken away all hope from some individuals and given it to millions of others. It was just possible that it, or something like it, was *the* cause of Parkinson's disease. Langston thought about the risks of working with MPTP for anybody and everybody who had used it. He thought in particular about the first man to synthesize MPPP and its by-product MPTP, Hoffman–La Roche chemist Albert Ziering. His formula had been developed in the 1940s. Ziering may well have worked with MPTP before anybody else. What had become of him, Langston wondered. Was he still alive? Was he MPTP's first victim? Perhaps he knew things which might help future research.

Langston, to whom detective work had now become second nature, decided to track him down. It took a little time to negotiate the layers of bureaucracy at Hoffman–La Roche, but eventually they gave him Dr. Ziering's home number. Now retired, Langston found him a bright and sparkling character who was fascinated by the MPTP story, and no, he didn't have Parkinson's disease. Ziering explained that his job had been to make new chemicals as potential drugs. After that, they were sent to other divisions for additional testing, and Ziering rarely heard the final results of his labors. But after they had been speaking for a while Ziering paused and said, "You know, I vaguely remember getting back a report that the compound was toxic."

Langston was stunned. "How could that be? Why would they have tested MPTP in the first place?" It was simply a by-product of MPPP synthesis.

Ziering laughed. "I really don't remember. But it may have been because of its general similarity to some anticholinergics; our company was interested in that class of compounds at the time."

Langston felt an uneasy sensation. In those days, before the advent of L-dopa, the only drugs available for treatment of Parkinson's disease were anticholinergics.

"So the next step would have been to test MPTP in animals?" Langston asked.

"Yes, I suppose so. But you would have to check back with the company to find out any further details. Sorry I can't help you more."

Langston thanked Ziering profusely, then immediately called Peter Sorter, the current chief of research for the company, to see if he could shed more light on the story.

Sorter did some research and eventually called Langston back. "I've found the report that Dr. Ziering must have been referring to," he began very deliberately. "It was filed in 1952. I can't send a copy to you, but I will read it over the phone." Langston wrote notes as Sorter spoke. Roche had tested MPTP on a series of animals. They first extensively tested MPTP on rats and found that it had no effect. Then they tested it on six monkeys. Thirty years before Stan Burns, they had seen that the monkeys froze up, became rigid, developed tremor, and died. But Sorter hadn't finished. According to a 1960 summary sheet, MPTP was also tested on people. Six humans, two of whom died during or after the study.

"Can you tell me what they were using it for?" Langston asked.

"Well, you're not going to believe it, but it was being tested as a possible treatment for Parkinson's disease."

In the next few months, Langston would learn that at least two other major pharmaceutical companies had discovered and tested MPTP-like compounds, years before he had seen George Carillo. In 1980, DuPont had tested an analog of MPTP (TMMP) as an antidepressant, but had abandoned it when they discovered it caused parkinsonism in monkeys. In 1983, shortly before Langston's *Science* paper came out, Eli Lilly was actively testing MPTP in animals as a possible anti-hypertension medication.

Ever since Ziering's discovery, MPTP had kept popping up. Pharmaceutical chemists had made it and tested it, designer-drug makers had synthesized it (unintentionally), and hundreds of drug addicts had consumed it long before it had come into Langston's life.

The strangest thing of all was that one company had actually been selling MPTP for years—the Aldrich Chemical Company. It was a minor compound, so minor that it hadn't been listed in the national database of chemical compounds, but it had some use as a chemical intermediate. Until the *Science* paper, however, orders for MPTP had been few and far between. The *Science* paper had changed that. Within hours, the phone had started ringing as scientists all over the world began to order the substance for experiments. Within days they had sold out and had to reorder new batches of the compound, which they sold both in a raw form, the so-called free-base MPTP, and in a slightly more expensive form as a hydrochloride salt. An obscure compound that hardly anyone wanted had suddenly become the hottest thing around.

When MPTP reappeared in the Aldrich catalog, it was more expensive. Over the next two years it would undergo spectacular price inflation as its unique properties were exploited for medical research. The price for 5 grams of MPTP went from eleven dollars in 1982 to ninety-five hundred dollars two years later. Langston liked to joke that a bottle of MPTP in his refrigerator (which Aldrich had given him for research in 1982) was now worth over two hundred thousand dollars, roughly five times what addicts on the street paid in 1982.

The Aldrich Chemical Company didn't actually manufacture the MPTP themselves; they subcontracted the work to a chemical manufacturing company based in Montreal.

By coincidence, Montreal was where Dr. Andre Barbeau lived and worked. Andre Barbeau was a living legend. His energy had no bounds. The author of over five hundred scientific papers and editor of twenty-eight books, he was one of the first physician-researchers to try L-dopa in humans in the early 1960s, and also among the first to recognize its terrible side effects. A very volatile and colorful Frenchman, he had been following the MPTP story with great interest.

Barbeau knew the Langstons. As part of her medical training, Langston's wife, Lisa, had done research on a hereditary neurodegenerative disorder known as Joseph's disease. As part of this effort, she had served on a research team with the ubiquitous Barbeau. During a dinner meeting in the summer of 1983 in New Bedford, Massachusetts, to which Bill Langston had tagged along, Barbeau reached into his overstuffed briefcase and whipped out a copy of the *Science* paper, which was dog-eared and crumpled from being carried around for months. "Just what does this mean?" he asked in an almost demanding voice. By the time coffee was served, Barbeau and Bill Langston had become deeply involved in a discussion of Parkinson's disease.

The issue which especially interested Andre was the etiology—the cause—of Parkinson's disease. If this simple chemical, MPTP, had caused parkinsonism in young California drug addicts, then perhaps other compounds in the environment might have a similar effect.

After talking to Langston and hearing about the other chemists who had developed symptoms, Barbeau started to think about the Montreal company which made MPTP. When Barbeau got interested in something, it was impossible to stop him. He asked to be allowed to visit the factory and to talk to the chemists and technicians responsible. When the bureaucrats at the chemical manufacturing company started to give him the runaround—they were prepared to let him visit the plant, but not to talk to the personnel—he lost his temper. Despite his protestations, they refused to let him meet the person who worked most actively with MPTP. Barbeau prepared to visit the plant anyway, and invited Bill Langston to join him.

The plant was a rather plain industrial building. They were met in the lobby by one of the executives, who gave a brief history of the company, which had changed hands recently, and a general description of the plant, including such things as the ventilation system and safety procedures. He denied that there had been any health problems among the employees, but did mention one "malcontent" who was constantly complaining of problems such as fatigue, stiffness, and malaise, which he assumed was his way of trying to get some type of disability compensation.

"What happened to him?" Barbeau asked.

"Oh, we finally transferred him to another area of the plant."

Next they were allowed a walk-through of the facility. As they walked along, Barbeau continued to pepper their host with questions. Suddenly, Barbeau stopped dead in his tracks. "What's the matter?" asked their anxious guide. "Is that man over there washing glassware the one that you transferred out of the MPTP area?" Barbeau asked, pointing to a lab technician in the far corner of the room. Langston gasped. There stood a bearded gentleman in a white coat who looked classically parkinsonian. He moved very slowly, with a flexed posture and short shuffling steps. He exhibited little facial expression and appeared to have aged far beyond his years. "Why, yes," answered their guide, his face turning almost white. "How on earth did you know?"

Long after the official had shown them off the premises, Langston and Barbeau continued talking about the implications of what they had seen. MPTP was a chemical that caused parkinsonism. How could this knowledge help scientists solve the greatest mystery of Parkinson's disease—to find its cause? Moreover, how could the discovery of MPTP help scientists find better treatments for the Parkinson's disease, and perhaps one day a cure?

13

New Treatments

MPTP was dangerous. It had maimed drug addicts, it had crippled scientists. But just why the brain's elaborate defense system had allowed it through the blood-brain barrier and into the dopamine neurons (and only the dopamine neurons), nobody knew. But from the fall of 1982 a series of dedicated scientists had struggled to find out.

From the start, Sandy Markey had recognized that MPTP itself was not the whole story. Back in the late summer of 1982, he had been immediately struck by the manner in which it affected Stan Burns's monkeys. Multiple doses and a fair amount of time were needed before the monkeys contracted the full-blown set of symptoms. This implied that MPTP was undergoing some changes—perhaps a series of changes—in the brain. It implied that there was some sequence of events. One possibility was that MPTP was being metabolized into another chemical.

Unaware that Irwin and Langston had formed a similar

hypothesis in their discussion at the Stanford cafeteria, Markey set about finding a way of tracking the compound after it was injected. This meant tracking where the molecules themselves ended up—no easy task. The MPTP molecule consisted of two ring-shaped structures, which would undergo change in metabolic reactions. Markey reasoned that the most stable part of this structure was the benzene ring—in most drug metabolism studies, these rings remain intact even though they are substituted in different molecules. If he tagged this ring, he could find out where, and in which molecule, it ended up. Markey used an old chemist's trick—radioactivity: he substituted radioactive carbon 14 for one of the carbon atoms in the ring. Now the MPTP was mildly radioactive and traceable.

The experiments proceeded methodically. The radioactively labeled MPTP was injected into rats, hamsters, and primates. When the monkeys developed parkinsonian symptoms as anticipated, they were sacrificed and their brain tissue was sent to Markey for chemical analysis. He separated out the chemicals in the brain tissue, carefully tracking the ones that had carried the radioactive tracer. He found that the major radioactive compound in the brain was no longer MPTP, but a slightly altered compound called 1-methyl-4-phenylpyridinium, or MPP+. As he had speculated, MPTP did not survive in the monkey's brain, but was transformed—metabolized—into something else. This MPP+ remained in the monkeys' brains for weeks or even months. Repeating the procedure with sacrificed rats and hamsters, Markey observed that in these unaffected species MPTP was also metabolized into MPP+, but they rapidly eliminated it from their brains.

Markey wondered whether MPP+ was the true culprit. If so, it would explain why primates were affected and rats weren't. Being oxidized, it could damage nerve cells. It was possible that in reality MPP+ was the toxic agent and MPTP itself was not toxic, but was what is known as a protoxin.

The next step was taken in San Francisco by a Japanese

graduate student from the University of Tokyo named Kan Chiba. In 1984, Chiba was on a fellowship at the University of California at San Francisco. His professors—Dr. Neal Castagnoli and Dr. Anthony Trevor—had asked him to help with a fascinating project to examine the metabolism of MPTP.

They were one of the hundred or so research teams now studying the newly discovered neurotoxin. They had been working with rats. Even though the animals didn't develop parkinsonian symptoms, their brain tissue still took up MPTP and their brain cells were affected. It was therefore possible to study the neurochemistry of MPTP using rats. The rats were sacrificed and their brains homogenized and subjected to the exquisitely precise methods of analytic chemistry. The team searched the homogenized brain tissue for chemical products and found one major substance.

Neal Castagnoli had known Sandy Markey for years and knew that Markey had been investigating the metabolism of MPTP in animal brains. After talking to Markey on the phone, Castagnoli wondered whether the substance they had found in the rat brains might be MPP+. Markey sent them some MPP+ for comparison in a mass spectrograph, and there was little doubt: the major substance in the rat brain tissue was not MPTP but MPP+.

Chiba was curious. If MPTP was converted into MPP+, how did it happen? Chiba thought that the agent was probably some very common chemical inside the brain—perhaps an enzyme whose function was to clear away the excess neurotransmitter. The enzyme he favored was called monoamine oxidase, or MAO.

Neurons communicate by a process that is part electrical, part chemical. The signal that passes along the axon is electrical. But when the electrical signal reaches the end of the axon, neurotransmitter molecules stored in tiny packages called vesicles are released. They then cross the tiny space—called the synaptic space—between the nerve terminal and the receiving portion (called the dendritic process) of the next nerve cell. If the molecules have the right shape, they bind to receptors on the surface of the second neuron. Once in place, the neurotransmitter alters the membrane so as to send a signal along the second neuron. Depending on the type of neurotransmitter and receptor, there are a wide variety of messages that can be sent and/or received.

Having achieved its goal and communicated, the neurotransmitter is then cleared out of the synaptic space so as not to interfere with future communications. This is achieved in various ways. Some neurotransmitter is reabsorbed by the membrane of the first cell, in a process called re-uptake. The rest is broken down by "janitor enzymes" into inactive small molecules. The common neurotransmitters like dopamine and norepinephrine are sometimes called monoamines, because in part they resemble a single ammonia molecule. The special enzyme that deactivates such monoamines is therefore referred to as monoamine oxidase.

One of the first classes of antidepressant drugs had been based on inhibiting the action of monoamine oxidase. Reasoning that depression resulted from a deficient supply of one or other of the main monoamines, scientists had set out to keep what monoamines there were in the synapse and to prevent their reabsorption and deactivation. Because these early antidepressants directly attacked the metabolism of monoamines by MAO, they were called MAO inhibitors.

Chiba's hypothesis went something like this: MPTP entered the brain where, because of its molecular shape, it was mistaken for the monamine dopamine and metabolized by the enzyme MAO into MPP+, which Markey and others were postulating was a neurotoxin. If this was happening, there was a very easy way to find out: use an MAO inhibitor to block the action of the monoamine oxidase enzyme. Without MAO, the MPTP would not be transformed into MPP+ and should be harmless. Working entirely in vitro (in the test tube), Chiba added two different MAO inhibitors—first deprenyl and second pargyline—to the suspension of MPTP-contaminated rat brain. If he was right, both pargyline and deprenyl would stop the action of the MAO enzyme and the MPTP would not be metabolized but would remain as MPTP. And indeed, when he analyzed the brain tissue all he found was MPTP. The reaction had been stopped in the test tube.

Chiba had made an inspired and important discovery. News of it spread rapidly through the scientific community. Castagnoli told Markey, who in turn told Richard Heikkila of Rutgers University Medical School, who told Langston. It would be months before Chiba's work would be published, and several scientists simply couldn't wait to try the experiment in vivo. Heikkila and Markey tried giving MPTP with an MAO inhibitor to mice. While MPTP-injected mice exhibit no clinical symptoms, the MPTP usually causes a decline in striatal dopamine. But Heikkila and Markey found no such dopamine decline: the MAO inhibitor had effectively blocked the transformation of MPTP into MPP+ in vivo.

Langston wanted to try out the idea on monkeys, where MPTP caused the clinical symptoms of parkinsonism. Together with his wife, Lisa, who was both an M.D. and a Ph.D., and was helping him in the lab until he could afford a technician, Langston crushed up pargyline tablets they had hurriedly got from the hospital pharmacy, mixed them with strawberry jam and marshmallows, and fed them to the monkeys. Then he gave the monkeys a dose of MPTP large enough to induce severe parkinsonism. The MPTP had no effect at all. The monkeys jumped around their cages, showing no signs of parkinsonism. More important, a later examination of the brain tissue showed that there was no death of cells in the substantia nigra. They were completely preserved.

Furthermore, when the brains were analyzed for MPTP, little or none was found. By early 1985 it was clear what was really happening. MPTP by itself is not toxic. Its toxicity depended on a naturally occurring brain enzyme which, by chance, converts it into a charged compound, MPP+, which by an even smaller chance is mistaken for dopamine and taken up into the neuron's membrane. This was the complex way MPTP caused cells in the substantia nigra to die.

Researchers realized that the finding had possible therapeutic consequences for patients with ordinary Parkinson's disease. Just as MPTP is metabolized into an oxidized substance (MPP+) by MAO, so too is dopamine. While stored dopamine is protected inside the neuron's storage vesicles, once released from the vesicles it is no longer protected. When a molecule of dopamine comes into contact with MAO, it is degraded, yielding a substance called 3,4-dihydroxyphenylacetic acid, or DOPAC, and hydrogen peroxide in the process. Having lost its ability to act as a neurotransmitter, the DOPAC is metabolized further into homovanillic acid (HVA), which can be measured in the cerebrospinal fluid (Burns's work at NIH had been concerned with measuring dopamine metabolites like HVA in the spinal fluid). Thus, not only is dopamine degraded, but potentially damaging hydrogen peroxide is produced as a side product.

Hydrogen peroxide is potentially dangerous because it can lead to the production of free radicals. Free radicals are molecular fragments that have one or more unpaired electrons and have net negative charge. Because of this they are highly reactive, being capable of rapid oxidizing reactions that destabilize or damage other molecules. Free radicals are widely used in the chemical industry precisely because of their ability to initiate and rapidly propagate chemical reactions. In the human body, however, and particularly in the brain, free radicals are capable of causing havoc, rupturing lipid membranes and destroying DNA and proteins and thus killing or maiming brain cells. Some scientists even believe that free radicals are directly involved in the aging process itself. For these reasons, free-radical scavengers such as vitamin E, which mop up free radicals, are promoted aggressively by many health magazines.

Thus, the reasoning went as follows: MAO oxidizes dopamine, producing hydrogen peroxide and associated free radicals, which kill neurons in the substantia nigra. If the oxidation of dopamine by MAO is damaging in this way, then blocking MAO (to slow dopamine's metabolism) might be healthy. In other words, treating patients with MAO inhibitors early in their Parkinson's disease *might slow the progress of the disease*.

It was an appealing idea which had already occurred to Walley Birkmayer, the Viennese physician who had first used L-dopa to treat Parkinson's disease in 1961. Before even hearing of MPTP, Birkmayer had reasoned that MAO inhibitors might be clinically beneficial to Parkinson's disease patients because they might help conserve what little dopamine was left. He reasoned that one could preserve the amount of dopamine at the synaptic cleft by blocking its degradation.

In 1983, Birkmayer and colleagues had reported on a study which compared a group of 323 Parkinson's disease patients who had been given L-dopa with a group of 285 Parkinson's disease patients who were receiving the MAO inhibitor deprenyl in addition to L-dopa. The original intent of the study was to see if deprenyl enhanced the effects of L-dopa, but in reviewing the data retrospectively, Birkmayer became suspicious that deprenyl was slowing the progression of Parkinson's disease. In 1985, he published a second study comparing a control group of 377 patients (on L-dopa alone) with 564 patients who received L-dopa plus deprenyl over a nine-year period. The deprenyl group lived on average fifteen months longer than the control group. Birkmayer's team interpreted these finding as suggesting deprenyl was actually preventing the death of nigral neurons in Parkinson's disease.

The stakes were very high. If a neuroprotective effect of deprenyl could be confirmed in controlled multicenter trials, then it would mean that the tragic progression of Parkinson's disease might be slowed down. Indeed, it would be the first time in history that any progressive neurodegenerative disease had been slowed. Birkmayer had used deprenyl with advanced Parkinson's patients, but Langston and other neurologists saw the possibility of a new therapeutic strategy, aimed at early intervention. If Parkinson's could be detected early and patients put on a suitable MAO inhibitor, then it might protect their cells and significantly slow the progression of their disease.

Of course, proving scientifically that deprenyl really did slow Parkinson's disease was a complicated exercise that would take several years. But for the first time in many years, there was new hope for Parkinson's disease patients.

Viewed from the standpoint of all this esoteric chemistry, the designer-drug disaster began to look ever more improbable. By chance, the chemist at the Morgan Hill lab had made a substance, MPTP, which could enter the brain freely, whereupon it was acted upon by a commonly occurring brain enzyme. Because of its precise chemical structure, it was converted into an electrically charged molecule, MPP+. But then something even more improbable happened. MPP+ just happens to fit the same re-uptake system which handles dopamine—it fits so well in fact that it is taken up by nigrostriated neurons as quickly as dopamine itself. Failing to recognize MPP+ as a lethal toxin, the brain delivers it to the critical neurons in the negrostriated system, where the charged compound becomes trapped and starts killing cells. Langston referred to this as the Trojan horse effect. From a scientific standpoint, it was precedent setting and if ever there was a cautionary tale about the downside of drug abuse, this was it.

After finding MPP+ in the brains of parkinsonian monkeys, Sandy Markey had done a literature search to see what was known about the compound. To his surprise, he noted that MPP+ had been developed some years earlier as an experimental herbicide called Cyperquat, chemically related to the much better known herbicide Paraquat. It had been test marketed in the early 1970s as a selective herbicide to clear land of weeds, especially nutsedge.

Cyperquat had undergone the usual toxicity tests in animals and the results were written up in the *Proceedings of the Twelfth British Weed Control Conference* in 1974. The tests showed that it was pretty toxic when ingested. When given orally to animals, it killed them by damaging their lungs and kidneys. Markey included reference to Cyperquat in his 1984 article on MPP+ for the prestigious scientific journal *Nature*.

Andre Barbeau also recognized the similarity of MPP+ to Paraquat and resolved to do his own investigation in Quebec aimed at finding the cause of Parkinson's disease. He began gathering data on the incidence of Parkinson's disease across the nine hydrographic regions of the province, and to everyone's surprise he discovered that, far from being uniform, it varied sevenfold across the different regions. It was highest in region three, the main agricultural area of the province, the so-called bread basket of Quebec. Next, Barbeau plotted the use of pesticides in Quebec, and the map was almost identical. Where there was pesticides, there was more—considerably more—Parkinson's disease than average.

Barbeau presented his preliminary results at the Ninth International Parkinson's Conference in New York in the summer of 1985. He was quick to point out that his findings didn't mean that pesticides were responsible, but rather that there was something in the agricultural landscape that was to blame. The data showed another striking correlation. The five hot spots that had by far the highest amounts of Parkinson's disease were all near pulp and paper mills—just why, Barbeau wasn't sure. Barbeau's results were controversial and many scientists were highly skeptical, but few were willing to dismiss the work of such a distinguished and innovative researcher. They waited to see where his new research program would lead. Tragically, the following year Barbeau suffered a heart attack and died.

By the summer of 1985, three years after the original tragedy, Parkinson's disease research had been transformed.

The public knew little about this revolution. The press had been more interested in the designer-drug story than in the medical research which it had spawned. For them, the frozen addicts were a cautionary tale which dramatically showed the evils of designer drugs. In TV news segments, Langston was called on to talk about the addicts' parkinsonism. In three years, Langston had appeared on dozens of TV shows and had appeared before congressional hearings on designer drugs. Although he would never relish such appearances, he was now a confident communicator.

He was not as confident about his newly independent career. Having been booted out of the seventh-floor office he had enjoyed as department chairman, he now operated out of half a trailer which he shared with the Mothers Milk Bank. Immediately behind the trailer were animal facilities where dogs were kept, so it got rather noisy at times. His research group also had available limited lab space at the Institute for Medical Research. The group consisted of Ian Irwin, Jim Tetrud, Lou DeLanney, Lysia Forno, and his wife, Lisa. Phil Ballard had finished his fellowship and moved to Seattle, where he went into private practice. The research group had now published a number of papers on the neurochemistry of MPTP—including articles describing the first identification of MPP+ and showing that one could block the effects of MPTP in primates—and were gaining a growing international reputation in the world of neuroscience. If Langston survived financially, his long-term prospects looked promising. Against all odds, the group had somehow managed not to be brushed aside by NIH, and were in the vanguard of scientific research.

Stan Burns had left NIH and taken a position at Vanderbilt University in Nashville, which had a large primate center. Sandy Markey and Irwin Kopin stayed at Bethesda, Markey remaining at NIMH to continue his research, Kopin becoming scientific director of the National Institute of Neurological Disorders and Stroke and a major figure in neuroscience research.

Langston knew there was very little hope for the six patients whose fatal encounter with MPTP had led to a breakthrough in Parkinson's disease research. The dopamine-making neurons in their brains were nearly gone, and the only possibility was to replace these cells with new ones—a brain tissue transplant. In 1985, this was still science fiction. There were early reports of monkey work at Emory University in Atlanta and at Yale, in which researchers had given monkeys MPTP parkinsonism and tried to reverse it by surgically implanting new brain cells. If successful, such experiments might one day be tried in humans. But this might take decades, and Langston was by no means sure that his six addicts would survive the rest of the 1980s.

Part Two

■ ■ ■ ■

14

Mending the Brain

More scientific knowledge has been amassed in the twentieth century than in all the centuries before it. Significant scientific findings are occurring with such frequency—there are over two thousand medical journals just in the United States, for example—that individual scientists can only keep up with a fraction of them. It was not surprising, therefore, that Langston didn't know very much about the pioneering work of Professor Anders Björklund and Dr. Olle Lindvall, and nothing about some operations which had taken place on Parkinson's disease patients in the spring of 1985.

Lund, Sweden, February 1985

On fine days, even in winter Anders Björklund liked to cycle the short distance from his home to the Department of Histology at Lund University. Short, thin, almost completely bald, he might have been taken for a country banker or accountant. In fact he was one of the finest scientists in the world. For fifteen years he had worked quietly in Lund pioneering a new frontier: brain tissue transplants.

Björklund was not, nor did he wish to be, a great communicator and popularizer. He was scrupulously polite, but he didn't especially enjoy talking about his work to the nonspecialist. He knew that it was not only difficult to understand, but also controversial. From time to time he had given popular lectures about his work, and he had found the questions very trying.

People would sometimes ask him about the possibility of doctors' one day performing "brain transplants." He hated the phrase "brain transplant." It conjured up an image of a mad scientist transplanting a whole brain from one individual to another. It was true that most other transplantations to date had involved whole organs—kidneys, livers, hearts, lungs, corneas. But the brain was different. The brain was not just an organ in a person's body; in a sense it *was* the person. If it were ever possible to take an entire brain out of one person's skull and transplant it into another body, then one would have effectively transplanted the person. More accurately, one might call such a procedure a whole-body transplant, because scientists would have in effect given a person a new body to go with his brain. All this was idle speculation, because that kind of procedure was strictly science fiction anyway.

Björklund had more modest and more realistic hopes. But they were no less exciting. In the early 1970s, he had become fascinated with the idea of repairing damaged brains. There were millions of people crippled with Alzheimer's, Parkinson's, and Huntington's disease, and countless numbers of individuals whose lives were plagued with epilepsy or ruined by strokes. Apart from some unsatisfactory drug treatments, nothing could be offered to these patients. In these diseases, brain cells were damaged and destroyed, and unlike the cells that make up skin, bone, or muscle, brain cells cannot be regenerated. In the brain what's lost is gone forever.

But Björklund wondered whether it might it be possible to transplant new brain tissue to replace the dead or damaged tissue. Several scientists before him had tried grafting brain cells from one animal to another. Most of these efforts had met with failure. The grafted brain tissue died. It was not immunologically rejected; it didn't survive long enough to be rejected. The fragile neurons simply perished for reasons that were not completely clear. Perhaps they died from oxygen deprivation; perhaps they were destroyed by mechanical handling.

Scientists had found only one exception to this rule: embryonic brain tissue. In 1940, at Oxford, Le Gros Clark found in rabbits that if he transplanted cells from the brain of a fetus into the brain of another rabbit, some of those cells survived. It was almost as if fetal neurons, beings immature, were more tolerant of being moved. With this starting point, Anders Björklund began a series of elegant experiments on rats, to chart the new frontier of neural transplantation.

Histology is the study of the structure of the tissues that make up living organisms. When Anders Björklund had started work, very little was known about the tissues that make up the human brain and precisely how they are damaged in neurodegenerative diseases. The problem was exacerbated by the fact that there were no animal models for any of the conditions. Animals didn't get Parkinson's disease, Alzheimer's, or Huntington's. Without animal models, scientific research is stymied.

So initially, everything depended on finding animal models—however imperfect—in which to study the diseases. Of all the neurodegenerative disorders, Björklund had thought that Parkinson's disease was the most promising. Working with rats, he had used a substance called 6-hydroxydopamine to destroy dopamine-making cells on one side of the rat's brain. Since this compound won't cross the blood-brain barrier, Björklund had injected it directly into the brain. Following injection, the rat displayed a dramatic change in behavior. When the rat was put in a confined space like a flat-bottomed bowl, it began to turn in circles toward the damaged side. Round and round it went. It could stop, but it was unable to rotate the other way.

This classic experiment, the rotometer experiment, had been repeated many times. It was in effect a crude model for Parkinson's disease. Björklund had directly damaged the dopamine system in a rat's brain and produced a very visible movement disorder. Moreover, it was possible to quantify the disorder by measuring how often the rat turned in a given time. The point, of course, wasn't to disable the rat, but to try and "cure" it of its "parkinsonism."

The cure consisted of a graft of cells from the fetus of another rat. The cells were taken from the substantia nigra of the rat fetal brain and transplanted into the striatum of the parkinsonian rat. The original destiny of these immature substantia nigra cells was to make dopamine. But would they make dopamine in another rat? After hundreds of such experiments, the answer was clearly yes. Parkinsonian rats that could only turn clockwise in the bowl would, after a brain tissue transplant, recover their ability to turn in both directions. The improvement was slow and steady, but over time the asymmetry disappeared.

Björklund had then gone on to prove beyond all possible doubt that the change was permanent and graft derived. When Björklund had surgically removed the graft, the unidirectional circling behavior returned. When he reapplied the graft, it went away. Microscopic analysis of hundreds of brains from rats which had received grafts revealed that the graft—the transplant—was surviving and sprouting nerve fibers. What surprised Björklund was the capacity of the nerve cells to re-create nerve connections with the surrounding host cells.

Anders Björklund parked his bike, ambled to his office, and began opening his mail. He was just ripping open his third envelope when first the head and then the tall, lanky figure of Patrik Brundin appeared in the doorway. "Sorry, Anders, are you busy?"

Björklund smiled. He was always pleased to see his young colleague. "Hello, Patrik, how is it going?"

Patrik grinned. "Great. I'm on my twentieth rat embryo this week. I wondered if you had heard anything from Olle Lindvall about the adrenal operations?"

"I haven't heard anything definite," Björklund replied. "Hold on a minute and I'll give him a call."

Patrik Brundin had gone to school in England and spoke English with a perfect Oxbridge accent. He had been drawn to this line of research because of personal circumstance. He had watched as his father slowly became crippled with Parkinson's disease. It had been a terrible tragedy for his family, but one which had given his life a focus. He had decided to go into medical research and search for a cure. He had come to work with Björklund and embarked on one of the great remaining scientific adventures. In the few years they had worked together, Brundin had quickly realized that Björklund was simply head and shoulders above many other neuroscientists, and they had become friends. If a cure was to be found, there was a very good chance that it would be found in Lund.

But Brundin often wondered whether Björklund's brilliant research would ever get out of the lab and be used on human patients, especially as it seemed to depend on fetal tissue. Even if the technique could be made to work in humans, it didn't mean that society would allow it. The history of organ transplantation had been racked with ethical controversy about the definition of death, about the regulation of living people donating kidneys. The use of fetal tissue raised questions that were even more ethically complex. What's more, it invoked the passions of abortion politics.

It was one thing to graft bits of the brain of a rat fetus into the brain of an adult rat; quite another to implant bits of the brain of a human fetus into the brain of an adult crippled with Parkinson's disease. Many would argue that it was fundamentally different from the usual kind of organ transplant, in which, say, the heart of an auto accident victim who is already dead is transplanted into a waiting recipient. Fetal-tissue transplants would use material from the brain of a human fetus, a fetus which might have become a person.

Because the ethical issues surrounding fetal-tissue transplants seemed so intractable, a radical alternative surgical approach to Parkinson's disease had been developed, and there were strong rumors that soon it was to be tried in Lund on two patients with Parkinson's disease. Björklund wasn't directly involved, but his colleague, neurologist Olle Lindvall, was a central figure.

The operation—an autotransplant—involved removing adrenal cells from a patient's own adrenal glands and transplanting them into his or her brain. The first two such adrenal transplants had been carried out in 1982, with very little publicity, by neuroscientist Lars Olson and neurosurgeon Erik-Olof Backlund from the Karolinska Institute in Stockholm.

While most cells outside the brain do not make dopamine, some body cells share a common heritage with nervous tissue. In the developing embryo, some cells descend from the same kind of precursor cells as nervous system cells. These cells, sometimes called paraneurons, produce neurotransmitters and can be stimulated to extend nerve fibers. The best-known such cells are found in the adrenal glands, located above the kidneys.

The innermost region of the adrenal gland is called the adrenal medulla. It contains a class of cells, the chromaffin cells, which produce dopamine. Lars Olson had reasoned that if these chromaffin cells were implanted into the brain, they might make dopamine and reverse the symptoms of Parkinson's disease. Olson and his colleagues had tried adrenal medulla grafts on rats and had successfully reversed some of the abnormal body movements, but there were doubts about how permanent the change was and whether the grafted cells survived. Nevertheless, the attractions of this approach in humans were several. First, there were few, if any, ethical problems. Second, there were no worries about rejection. This was an autograft—a transplant within the body—and it would not, by definition, be rejected. Third, there was no problem finding the tissue; almost everyone has two adrenal glands—one for each kidney—and can survive perfectly well with one of them.

For Backlund, a devout Christian, the ethical arguments were critical. It was he who had gone to Lars Olson in 1982 suggesting that they try adrenal transplants for two patients whose parkinsonian symptoms could no longer be treated with drugs. As a neurosurgeon he was devoted to helping people with damaged brains, but he was not prepared to use human fetuses for this purpose. So, in 1982, he had surgically implanted adrenal cells into the caudate nucleus—one of the two areas that make up the striatum—in one side of each patient's brain. Unfortunately, the patients had shown only a minor improvement for a couple of months. Six months after surgery, there were no discernible effects. Two years after implantation, it appeared that the transplant had not affected the course of their Parkinson's disease at all.

Backlund had chosen the caudate nucleus site for a number of reasons. Logically, one ought to place the adrenal cells in the substantia nigra region of the patient's brain. After all, this is the area where the cells have died. This is the "input" part of the nigrostriatal dopaminergic system. In a normal brain, the dopamine-making cells of the substantia nigra extend axons several centimeters long to connect with targets in the striatum (the "output" side of the system). From here the effects of dopamine are orchestrated throughout the rest of the brain.

But dozens of animal experiments had shown that grafted nerve fibers do not grow such distances in adult brains. Cells grafted to the substantia nigra of rats did not reverse their Parkinson-like symptoms. In contrast, grafts placed close to the target—in or just outside the striatum itself—worked spectacularly well in rats. That was why they had chosen the caudate for the first two human adrenal transplants.

Now, three years out, the first two operations still did not seem to have made much difference to the patients. There were many possible reasons why. It might be the tissue itself—that adrenal tissue did not belong in the brain and therefore could not be expected to behave like brain tissue. But it might be the placement of the graft. There was some evidence that a better effect might be gained by placing the transplant in the nearby putamen, which makes up the other half of the striatum. Or it might be that any improvement in the patients had been so subtle that the doctors had missed it.

Because Backlund was a surgeon, and Olson a basic scientist, neither was ideally qualified to assess the neurological effects, if any, of the adrenal transplant operations. Backlund had needed a highly qualified and methodical neurologist to carry out the long and subtle assessments in these and subsequent patients. Remarkably, he had been unable to find a neurologist in Stockholm who was both qualified and interested in doing this assessment. So he had recently turned to Lund to discuss his results with Anders Björklund's colleague, neurologist Olle Lindvall.

Olle Lindvall was very interested in helping. A clinical neurologist with a background in basic research, Lindvall was perfectly qualified to determine whether the new operation worked or not. Lindvall offered to select two new patients in Lund who would be offered an opportunity for an adrenal transplant. Backlund would do the surgery and he would carry out a thorough neurological assessment to determine the outcome.

Anders Björklund put down the telephone and turned to Patrik Brundin. "Olle Lindvall says the operations are set for sometime in March."

Hospital de Especialidades Centro Medico "La Raza," Mexico City, Spring 1985

Dr. Ignacio Madrazo told his patient to stand up and walk back and forth in the clinic. Jose Luis Mesa walked dragging his right foot behind him, like a lame old man. He was hunched over and shuffled in the characteristic parkinsonian manner.

"Okay, Señor Mesa, pare."

Señor Mesa stood facing Dr. Madrazo. His face had no expression, his right arm had a pronounced resting tremor. Madrazo had not seen many cases as severe as this one. The factor which made it especially interesting was that Señor Mesa was only 35 years old.

Until five years before, Jose Luis Mesa had worked for the Mexican Railroad Company. Then he had come down with young-onset Parkinson's disease—a rare but well-documented condition that affected a small percentage of the population. It had come on quickly and progressed rapidly. Now he was an invalid. Not only could he not work, he could not feed himself or dress himself; he even needed help going to the bathroom. To make things worse, he seemed unable to tolerate any medication. L-dopa produced such terrible side effects that his doctors did not allow him to take it.

Mesa had been driven to seek help at the Hospital de Especialidades Centro Medico "La Raza" in Mexico City, where he had heard they were considering a new kind of operation for Parkinson's disease. He had been pushing them to take him on.

For several years, Dr. Madrazo and his colleague, Dr. Rene Drucker-Colin, had been doing laboratory work in animals, investigating the possibilities of adrenal medulla transplants. In Mexico, where abortion was illegal, fetal transplants were almost unthinkable. But Madrazo felt encouraged about the possibilities of adrenal grafts.

He had been waiting for a candidate to try out the technique on. In Señor Mesa he had found him: a young, very ill man with very little to lose and everything to gain.

Operating Room, University of Lund, March 1985

In March of 1985, general surgeon Dr. Anders Nobin opened up the abdomen of a 65-year-old man with advanced Parkinson's disease. The procedure went smoothly. After the abdomen was fully exposed, Nobin removed one of the patient's adrenal glands. Åke Seiger, a scientist from Stockholm and a close collaborator of Lars Olson's, discarded the gland's outer cortex, revealing the inner adrenal medulla. He then cut this into small slices and placed the tissue into a small cylindrical container, like a hypodermic needle ready for implantation.

The second part of the operation involved getting the pieces of tissue into the correct part of the brain, and would be carried out by neurosurgeon Erik-Olof Backlund. There are two main surgical approaches to implanting tissue into the brain. One is to remove a piece of skull, make a small channel through the brain to the ventricle, allowing the surgeon to look down on the caudate nucleus. The other is a less invasive procedure known as stereotaxis, which is done blind through a single small hole in the skull. Backlund elected to use stereotactic surgery, using a device developed by another Swedish neurosurgeon, Lars Leksell.

Imagine trying to target a submarine hidden deep under water armed only with the submarine's depth, latitude, and longitude. If all three are known with sufficient precision, then, given the proper aiming device, the submarine can be hit, even though it can't be seen from above the surface at all. A similar principle is involved in stereotactic surgery. Stereotactic surgery seeks to target a specific brain region that cannot be seen, by determining the target region's precise spatial coordinates.

The process is tedious and time-consuming. First a hemispherical frame is screwed into the head. This frame has coordinates marked out in three perpendicular axes. If one knows the right coordinates one can, by using the frame, drill a small hole in the skull and place the tip of a needle almost anywhere in the brain that is safe to go.

Earlier that morning, the team had bolted the stereotactic frame to the patient's skull. With this in place, they wheeled the patient to radiology for a CAT scan to reveal in three dimensions where the major brain structures were located. Using the scan, Backlund plotted the best approach into the brain, avoiding all major blood vessels, and a computer calculated the exact angles at which he had to enter the brain and the correct depth to deposit graft material.

Back in the OR, Backlund then set the coordinates on the stereotactic frame, drilled a small burr hole in the skull, and inserted the cannula to the correct depth, thereby "blindly" reaching the target location in the striatum.

Backlund began the long job of inserting the tissue. Once the tip of the cannula had reached its calculated target area, he withdrew the wire that filled its barrel, and injected a suspension of fragments of adrenal tissue and implanted it deep into the patient's brain, depositing the cells in the patient's right putamen.

Hospital de Especialidades Centro Medico "La Raza,"
Mexico City, Spring 1986

The operation was going well. The abdominal surgeon had removed one of Jose Luis Mesa's adrenal glands and the tissue had been cut into pieces, ready for insertion. Because Madrazo did not have access to the equipment necessary for stereotactic surgery, he had elected to use an open surgical technique. Madrazo knew the procedure well; the open surgical technique was frequently used in Mexico to remove intraventricular parasitic cysts. Using a special bone saw, Madrazo carefully removed some of the bone over the right frontal lobe of Mesa's brain. With the brain exposed, Madrazo passed an ultrasound probe over the surface of the brain itself. On a screen a picture of the reflected sound waves revealed the structures deep within the brain.

The different tissue densities gave a different echo and, like a pregnant woman's sonogram, revealed what was under the surface. To Madrazo's trained eye, the picture looked familiar. He was searching for the ventricle, the major fluid-filled cavity, and the striatum, which lay just below it.

Having located where he wanted to place the graft, Madrazo began opening up the brain. He made a narrow channel down to the ventricle. Once in the ventricle, he could see the caudate nucleus, as it bulges directly into the lateral ventrical of the brain. Then he carefully carved a cavity in it, large enough to fit the adrenal tissue.

Now everything was ready for the grafting. One by one, Madrazo placed the pieces of adrenal medulla into the specially cut cavity. This was the key to the open surgical technique. True, it was more invasive than stereotactic surgery. But it meant that the transplanted tissue would be directly in contact with the cerebrospinal fluid of the main ventricle. This fluid might nourish the cells and promote their production of dopamine. There was also some evidence from the University of California at Irvine that creating the cavity released special molecules called growth (or trophic) factors, which could stimulate the growth of new tissue. Madrazo was certain that the first Swedish operations in Karolinska had failed because they had used the stereotactic technique. Starved of oxygen and nutrients, the implanted cells had died.

Madrazo looked across the OR at his colleague, Dr. Drucker-Colin. It was an exciting time for neuroscientists and neurosurgeons, and especially exciting that Mexico was in the forefront of this research.

Lund, Spring 1986

Dr. Olle Lindvall finished examining the second of the two patients given adrenal transplants the year before in Lund. It had been a very thorough examination. Lindvall was nothing if not thorough.

The whole process had been a textbook exercise in methodical research and correct clinical practice. First of all, the team had been scrupulously careful in selecting their patients. The first potential source of error would be to choose a patient who didn't really have Parkinson's disease. The second source of error was controlling for all the variability in the patient's symptoms. Parkinson's disease patients were notorious for having spontaneous fluctuations in their symptoms. Then there was the placebo effect. Lindvall knew that this was especially significant in Parkinson's disease cases. Just entering a treatment research program led some patients to claim they felt better. Most important was the issue of medication. Unless this was controlled for, they would never know for sure what difference, if any, the graft was making.

Although he was primarily a clinician, Lindvall had a strong scientific background. Both he and Björklund felt that unless one proceeded scientifically, nothing would be learned from these experimental operations. So for six months prior to surgery and twelve months afterward, Lindvall had insisted that the two patients maintain exactly the same level of L-dopa medication. If their medication was varied in any way, it would be impossible to accurately gauge the effect of a brain graft.

Next, Lindvall and his team standardized the clinical examination the patients received during this critical year and a half. Lindvall wanted to make the typical tests that neurologists performed on Parkinson's disease patients to test their motor skills—walking up and down, pronating-supinating (flipping) their hands back and forth, and so on—as objective as possible. So week in and week out, the same battery of tests were performed and the performance was measured. A quantity—for example, the time taken to make a set of movements, or the number of hand flips in a given time interval—was recorded. These quantitative evaluations were done at standardized times—for example, early in the morning, before the patient had taken his first dose of L-dopa.

The clinical data had to be unambiguous and reliable, because it constituted the only evidence of success or failure.

Lindvall's colleague Anders Björklund had been able to "sacrifice" his rats and examine their brains to see if the graft had survived, but this would not be possible with the human patients. There were other, noninvasive, techniques just appearing on the scene, like the PET scan. By comparing the scan before and after the operation, it was possible to make measurements which indicated the degree to which the graft had taken. But there was still some controversy about the interpretation of these scans.

Now, having gone over all of the clinical data, Lindvall was completely convinced. The adrenal transplants had failed to produce any long-term benefit. At first it had looked promising. The first Lund patient, a 65-year-old man, had been able to get up and walk in the morning before taking his medication. The second Lund patient, a 45-year-old man, was able to care for himself most of the time. Both men had fewer and shorter off periods (periods when the medication wasn't working). But it hadn't lasted. The small improvements seen in the first months had now completely disappeared. Lindvall had effectively replicated what had been seen with the first two adrenal transplant cases at the Karolinska Institute in 1982.

The results were disappointing but not surprising. They confirmed what countless experiments on rats had shown: that adrenal grafts produced a small improvement following surgery, which soon disappeared. When these animals were autopsied a year after grafting, it was difficult to find any surviving adrenal cells, and there was almost no fiber outgrowth from the implanted cells. Putting adrenal cells into the brains of animals had not made them behave like brain cells. Like fish out of water, the transplanted adrenal cells had died.

There could be no question of doing any more human cases until some new strategy was developed. Adrenal transplants might be a dead end. There was of course another option: to use fetal brain cells. The rat data on fetal transplants was impressive. Vast numbers of rat experiments showed that fetal grafts survived for long periods and formed connections with the surrounding brain tissue. Scientifically speaking, it made sense to continue this work and try it in human cases. But would it be ethically acceptable to the public?

The Fetal-Tissue Transplant Task Force

After long discussions, Björklund and Lindvall decided that the most responsible way to proceed was to set up a special task force at Lund to explore the entire area of fetal-tissue transplants in humans. This would be done in collaboration with Lars Olson and Åke Seiger of the Stockholm group and with a group in Uppsala doing research on transplants of fetal insulin-producing cells in order to cure diabetes. A long and difficult road lay ahead. One goal of the task force would be to explore the ethical questions. To this end, they had decided to ask the Swedish Society of Medicine to initiate a public debate on the ethics of fetal tissue transplants. But there were a whole host of scientific questions that needed to be answered—questions about the amount of fetal material and where to place it; questions about the immunology of the brain. The task force would lay out a plan of the experiments that needed to be done and then set about doing them.

Olle Lindvall had heard of some exciting new work in America, on monkeys with MPTP parkinsonism. Researchers in Atlanta and at Yale had given monkeys MPTP parkinsonism—an animal model that actually replicated many of the physical signs of parkinsonism seen in humans—and had set out to reverse it with cells from the brain of a monkey fetus. The early reports were very exciting. These grafts had reversed virtually all of the motor signs of Parkinson's disease: tremor, rigidity, slowness of movement. And postmortem examination of the brains had shown that the graft had indeed survived and made connections with surrounding tissue. If it turned out that fetal tissue offered the only immediate hope of "curing" Parkinson's disease, different countries around the world would have to figure out what they wanted to do about it. Because the issue was so explosive, Lindvall and Björklund wanted to proceed cautiously and responsibly. Without public support in Sweden there would be no point in going ahead.

But midway through their planning, something happened which threatened to change the direction of neuroscience and upset their plans. And it came from Mexico City.

15

Miracles

Schmitt Symposium on Transplantation into the Mammalian Brain, Rochester, New York: July 1987

Dr. Rene Drucker-Colin walked to the podium to begin his address. The hall was packed with neuroscientists, clinical neurologists, and neurosurgeons, dozens of journalists, and TV crews. Many had brought copies of the April 2 issue of *The New England Journal of Medicine,* which had carried a report of his and Dr. Madrazo's first two adrenal-tissue transplants. Breaking with more than a century of tradition, the journal had bypassed the normal protracted review process and published their paper quickly. The editor, Arnold Relman, had thought it of such significance that it needed to be reported as soon as possible.

Founded in Boston in 1812, *The New England Journal of Medicine* is the oldest continually published medical journal in the world. It is also by far the most prestigious, enjoying the highest circulation of any professional medical journal. The *NEJM* prides itself on two things above all others. First, it publishes cutting-edge biomedical research papers. Second, those papers have to survive the *NEJM*'s rigorous and protracted peer-review process—considered to be as strenuous as that of any research journal. This review process, which takes up to a year, is a considerable source of tension to researchers who want to tell the world about an important new discovery. Yet, as the *NEJM*'s reputation depends more on the quality of papers it publishes than on the speed of publication, it rarely compromises this established procedure.

A handful of times in its long history, the editorial staff of the *NEJM* has decided that a new paper had such potential importance that it warranted special treatment. This is what the *NEJM*'s thirty-seventh editor, Arnold Relman, decided on seeing the Madrazo and Drucker-Colin adrenal transplant paper. Such is the influence of *The New England Journal of Medicine* that its reports are picked up by newspapers and TV stations around the world. Within days, Madrazo and Drucker-Colin had become celebrities. Their work was shown on TV and featured in *Time* and *Newsweek*.

Drucker-Colin and Madrazo were undoubtedly the star act at the Schmitt Symposium in Rochester, their presence enlivening a normally conservative scientific meeting. But had they really transformed neuroscience and found a cure for Parkinson's disease? The conversations in the hallways showed that the scientific community was divided. Many neurosurgeons were very excited by what they had done and were eager to try it with their own patients, but some of the neuroscientists doing basic research were openly skeptical. The Mexican findings did not fit with a lot of the basic research done in Sweden, and many scientists would not be satisfied until the work had been replicated in a major center in the United States or Western Europe.

As a scientist, Drucker-Colin thought this was fine. He certainly wanted to follow up a group of patients for at least three years before offering the operation to other Parkinson's disease patients. But he wondered how different things would have been if a group of Swedish rather than Mexican scientists had announced this breakthrough.

Today he had the opportunity to show in more detail what he and Dr. Madrazo had accomplished. He intended to illustrate his formal presentation with extraordinary videotapes of the patients before and after surgery.

His star patient was Jose Luis Mesa. He began by showing footage of Señor Mesa before surgery. The audience saw the hunched figure of a terribly disabled man with a severe tremor and a flat, expressionless face. Drucker-Colin described how until five years before, Jose Luis Mesa had worked for the Mexican Railroad Company. Then he had come down with young-onset Parkinson's disease and soon was an invalid, unable to work, feed or dress himself, or go to the bathroom unassisted. He could not take L-dopa because of the terrible side effects it produced in him.

Drucker-Colin then described the surgical procedure he and Madrazo had carried out at the Hospital de Especialidades Centro Medico "La Raza," in Mexico City. "We carried out unilateral surgery grafting adrenal medulla cells to the caudate nucleus. After fifteen to twenty days we could see a change. The first improvement was in his face, expression returned to it. Slowly, his other problems began to improve. This videotape was taken a few months after surgery." Drucker-Colin glanced at the audience. Their eyes were now riveted to the screen watching Mesa, who in the previous tape segment could hardly move, now walking quickly down a corridor—albeit with a limp. He heard murmurs of astonishment—"Wow" and "Take a look at that"—as the audience began to succumb to the magic of video.

Drucker-Colin saved the best for last. "This videotape was taken in June, and you can see he is much better, although he still has a slight tremor." The picture cut to Mesa's little farm. On the screen, he looked like an active outdoorsman, walking robustly, shoveling manure, even picking up a soccer ball and kicking it. The audience watched, spellbound. Surely this could not be the same man who was crippled at age 35. Drucker-Colin explained that Señor Mesa had returned to work and that he used this little farm to grow vegetables. He finished up. "We are continuing to follow him. This patient could not tolerate any medication before surgery and is on no medication now. Thank you very much for your attention."

Members of the audience applauded as they mulled over the implications of Drucker-Colin's presentation. The neurologists in the audience had already felt pressure from some of their critically ill patients to give them an adrenal transplant. When these videotapes were shown on television, the pressure would increase. The neurosurgeons in the audience saw an important new procedure for them to carry out. The press was excited by the tapes. Rarely were scientific conferences so interesting. But the basic scientists sitting in the audience were not ecstatic. They were puzzled—especially Anders Björklund and Patrik Brundin from Lund.

Björklund did not know what to make of the tapes. Ever since he had read the Mexican paper in *The New England Journal of Medicine,* he had felt doubtful. As a scientific paper, Anders Björklund had thought it was dreadful. Björklund could not imagine why Relman had agreed to publish it. The paper had not fully detailed the research methods, nor included enough data. Did the patients really have Parkinson's disease? What methods had been used to evaluate the severity of their Parkinson's disease? Then there was the pattern of their improvement. Madrazo's patients had experienced no improvement for the first two to three weeks, but thereafter the improvement had continued gradually over a year or more. There had been no falloff; the improvement seemed to be long-lasting. This was the opposite of what had happened in the animal experiments and in the four Swedish patients, where the improvement had been immediate but temporary, falling off over two to six months. A vast number of animal experiments showed that grafted adrenal cells gradually died off, with the cells becoming less and less functional. How could all these experiments be wrong? When results in humans didn't agree at all with a wealth of data built up in animal models, it could mean one of two things. Either the animal model was a bad model, or the human study was flawed.

Also, there was the fact that a unilateral transplant—a transplant on one side of the brain—had produced a bilateral effect—an improvement on both sides of the body. This was not impossible, as the two sides of the brain are connected, but it certainly required an explanation.

It just didn't make sense. As far as Björklund could see, the Mexican team's results were an effect in search of an explanation. But he didn't want to publicly attack their findings, and he worried that people might think that his reservations were motivated by professional jealousy.

Patrik Brundin was also feeling uncomfortable. He had agreed to sit on a panel that would discuss the current state of neural transplantation research and would take questions from the press. He would be sitting next to Madrazo and Drucker-Colin and would certainly be asked what he thought of their work. Also on the panel would be neurosurgeon George Allen, from Vanderbilt University, where Burns had gone after leaving NIH. Allen and Burns were now working together and were quite enthusiastic about the adrenal operation. Eight weeks earlier they had done their first case. While it was too early to form conclusions on this case, they had plans to do several more. Brundin had no idea what he was going to say.

Press Conference

Journalists attending the conference had discovered from conversations with scientists that this new adrenal operation had been carried out not only in Mexico but also in China, Cuba, England, and the United States, and while none of these groups had published any outcomes, it was clear that none had obtained results as dramatic as the Mexican team. The more seasoned science correspondents realized they were witnessing the start of a stampede. Pressure from patients, combined with the ego of U.S. surgeons, would ensure that over the next few months hundreds of these operations would be done.

Some of the more senior correspondents had seen a similar thing happen with heart transplants. After Christiaan Barnard's operation in Cape Town, South Africa, on patient Louis Washkansky in December 1967, American heart surgeons had gone crazy doing hundreds of heart transplants over the next two years. Knowing very little about the subtleties of immunosuppression, the surgeons had been helpless when the newly transplanted organs had been rejected. Virtually all of the patients had died. At the time, the careful Stanford surgeon and researcher Norman Shumway had warned against going too fast, arguing that you couldn't cheat science. He had been right. Only after twenty years of research did heart transplants become a routine procedure. The improvement in survival rate had come through a meticulous scientific approach, not by performing more operations.

In Barnard's day, there were not many science correspondents. Press coverage was naive and sensationalist. By 1987, there were hundreds of science correspondents around the world, many highly intelligent and sophisticated individuals. Several of those present at Rochester were fully aware that the Mexican results contradicted more than a decade of Swedish research. One of them directed a question to Patrik Brundin.

"Dr. Brundin, how do you explain the very different results the Mexicans have obtained?"

Patrik Brundin walked up to the podium. "I describe them as being different," he said, and gave them a broad grin. Nobody laughed. The group of reporters looked straight at him, waiting for him to answer the question properly. Nervously, Brundin appealed to the questioner, "No, seriously, are you asking why are they different?"

The reporter repeated the question more explicitly. "As an observer, how do you interpret the results they have presented?"

Brundin tried to be as diplomatic as he could, especially as Drs. Madrazo and Drucker-Colin were sitting a few feet behind him. "I think one should be cautious and see what the future will tell us. It's too early. I'm a bit puzzled and surprised, but I see the same film as you do and the patients definitely look better."

Another reporter chimed in. "Surprised about what?"

"I'm just saying that the rat data wasn't as encouraging as the clinical data. But I see the same films as you do and the patients definitely change."

Brundin looked at the bank of journalists staring at him. They didn't look satisfied. He wasn't satisfied. The truth was, he didn't know what to think, he didn't know how to explain the differences. But the surgical technique was different—that was something definite which he could talk about.

"There's a difference in technique which should be emphasized. In the Mexican procedure the tissue is in contact with the cerebrospinal fluid all the time. All our work was done using a stereotactic technique. Adrenal cells are responsive to trophic factors, and this may explain it—although I should add that experiments in rodents have been done leaving the tissue in the ventricle in contact with the CSF and they haven't shown much effect. But again, the rodent might be different than humans."

Brundin sat down, rather embarrassed. Immediately, Madrazo got up to reply.

"The greatest difference between the results so far might be time. By now our patients have had almost seventeen months' evolution of their graft. The other [groups that have carried out adrenal grafts have had less time]—in China it is only six months and in Vanderbilt it is only two or three months. Our patients have been improving all the time and we can expect theirs to also."

Brundin listened, far from convinced. It was now five years since the first adrenal cases done at the Karolinska Institute. In those cases, time had made no difference.

Madrazo continued. "I think our adrenal operation is a

very elegant procedure. It has no immunological problems and patients do pretty well. It seems to me that it is hard to make patients understand why we have to go to fetal transplants if adrenal transplants are doing so well."

Finally, Drucker-Colin got up. "In my mind, after doing so many cases, there is no doubt that they get better. The question is, for how long."

16

Nemesis

Vanderbilt University, Nashville, Tennessee, 1988

Neurologist Stan Burns turned on the camcorder and began his clinical examination. The patient, a 35-year-old man with juvenile Parkinson's disease, had been one of the first U.S. cases to receive an adrenal tissue transplant. Methodically, Burns ran through the standard set of tests. Everything was being recorded. To avoid bias, the videotape would later be "scored" by an independent scientist. The patient's head was covered, so that the scorer would not see whether it had been shaved for surgery—in other words, the scorer would not know whether he was seeing a patient before or after surgery. The scorer would objectively rate how the patient performed on a series of standard scales.

Burns had few regrets about leaving NIH for Vanderbilt. Vanderbilt had a large primate facility for basic research and was a much less political place to do research. But following the announcement of Madrazo and Drucker-Colin, Burns's senior colleague, neurosurgeon George Allen, had become very interested in setting up an adrenal transplant program at Vanderbilt. Burns and Allen had traveled down to Mexico City to study the procedure so that they could replicate it in a series of safe, controlled procedures at Vanderbilt.

It had been an interesting visit. On the one hand, the center seemed professionally run and the staff competent. The neurologists seemed capable of distinguishing Parkinson's disease from non–Parkinson's disease, and the patients certainly seemed to be better than they were before surgery (as evidenced by videotapes). On the other hand, the record keeping on the patients seemed quite inadequate.

Despite their abilities, Burns discovered that Madrazo's team had no experience with standard rating scales. Neurologists in the United States and many European countries used standard scales to "rate" their patients, such as the Modified Columbia Rating Scale or the Unified Parkinson's Disease Rating Scale. By using these scales, which measured a patient's performance on specific tasks, neurologists generated objective data that their colleagues could understand. But Madrazo and his colleagues seemed not to have heard of these scales. At La Raza, whenever a question was asked about the degree of improvement, Madrazo typically turned to the patient and asked, "How have you improved?" The patient then gave a personal history. This personal history was supported with impressive videotapes.

The videotapes looked dramatic, but they were more like home movies than clinical videos. Instead of standardizing the task so that like could be compared with like, the patients were filmed at different stages of treatment doing different things.

Then there was the diagnosis. While in Burns's view most of the patients had genuine Parkinson's disease or at least parkinsonism, at least one patient he examined there did not, in his view, have full-blown parkinsonism.

Despite these irregularities, Burns and Allen thought that the work deserved to be replicated under controlled conditions in Nashville. And when they got back, they started recruiting patients.

Parkinson's is a highly variable disease. While year after year, it progresses relentlessly, response to treatment can vary tremendously from day to day, and even month to month, often for no apparent reason. This variability makes the evaluation of new treatments extremely difficult. Simply by choosing to videotape patients at certain times, one can create an impression of dramatic and long-lasting improvement, when in reality there is none. To be certain that patients were in fact better and that their improvement was due to an adrenal graft, they needed to be followed for a long period with repeated, objective, and quantitative rating scales.

Experienced neurologists know that when you examine patients over a long period of time, you see them at their best and their worst. You see them when they are sick, or after a terrible night's sleep. Due to such common variables as these, sometimes a Parkinson's patient appears better, sometimes worse. One cannot read too much into an individual examination. Only the trend over a long period will convincingly demonstrate whether or not a transplant has worked.

The problem was that while Burns and Allen carefully recruited patients for their scientific study, others in the medical world seemed to have no qualms about plunging ahead. Following the Rochester meeting, surgeons all over the world had raced to do adrenal tissue transplants. Thus far they had been performed in Florida, Texas, Chicago, Los Angeles, Poland, China, England, and Cuba. Some surgeons had no scientific protocol at all and were simply responding to patient demand. Some groups had very poor experimental designs. Other groups had good designs but the designs varied so greatly that comparing the results between different groups would be impossible. In the twelve months following the Rochester conference, some three hundred adrenal transplant operations were done. In Burns's view, this was madness. What was the point of so many centers doing this operation in completely different ways with different patients and different protocols? What could possibly be learned?

At Vanderbilt their strategy had been to select three groups of patients. The first six transplants were given to patients under 50 with mild to moderate disease. The second set of transplants went to six patients under 50 with moderate to severe symptoms. The third set of transplants were performed on six patients between the ages of 60 and 70 with moderate to severe Parkinson's disease.

They had followed up the cases methodically for over a year, and had found nothing like the phenomenal results that Madrazo and Drucker-Colin claimed to have achieved. The older group of patients had experienced no sustained improvement. But for the twelve patients under 50, the results were at first mildly encouraging. After the first six months, nine patients scored the same or higher than before surgery. But then Burns observed that some of the gains had begun to erode. After eighteen months, the patients' Parkinson's disease began to get worse.

One of the patients had suffered a heart attack and died, from causes unrelated to the surgery, a tragedy that offered an opportunity to see whether the transplanted cells were surviving. A postmortem examination of his brain tissue revealed that some three months after transplant, virtually all of the adrenal cells were dead.

The patients at Nashville had been carefully screened before going ahead with surgery. Abdominal surgery combined with a craniotomy is a major procedure for anyone, let alone a 70-year-old. Such surgery always carries a risk of morbidity and mortality. But many groups were not as careful, and some patients had died of surgical complications, including a few U.S. patients who traveled to Mexico City for the operation. Some of these cases who had died from surgical complications were studied postmortem, revealing that their transplanted tissue was dead. The grafts had not survived.

If the graft wasn't surviving, then it was difficult to see why any improvement should be expected. Some scientists, like Don Gash of Rochester, argued that the benefit the Mexicans had observed might have nothing to do with the transplanted cells—the adrenal tissue was irrelevant; it was the surgery itself that had caused the improvement. Drilling a hole and making a cavity leaves the brain with a wound. Macrophages, a type of white blood cell which clean up debris at injured sites, had been observed at the surgical transplant site. Macrophages are known to release growth factors which might rejuvenate neurons that have been weakened by Parkinson's disease. This might account for any temporary improvement following surgery.

Madrazo's and Drucker-Colin's tapes had generated enormous hope and expectation. Now came disappointment and disillusionment.

The negative publicity following the deaths had a chilling effect. Patients who had demanded the operation now refused to volunteer for trials. Critics argued that the benefits were minimal and the risks substantial. Like stockbrokers in a bear market, the medical world decided to dump adrenal transplants as quickly as it had adopted them. Even though many had participated in the stampede, the media singled out Madrazo and his colleagues for blame. ABC's *20-20* did a critical piece accusing Madrazo of unethical human experimentation.

Burns and Allen had plans to do a large two-hundred-patient study to try and tease out whether adrenal transplants did have benefit for a small subset of patients. But as the bad results poured in from groups around the world, enthusiasm for the operation quickly waned. Eventually, they found it impossible to recruit patients, and by 1990 they had abandoned the study. Adrenal tissue transplants were effectively dead.

Yale–St. Kitts Biomedical Research Foundation, St. Kitts, Caribbean, Winter 1986

Dr. Eugene Redmond stood looking at the monkey in its cage. It was incredible. Just three months before, the African green monkey had been parkinsonian: immobile, frozen, mute, unable to feed itself following injection with MPTP. But now it was virtually normal. It jumped around the cage making a lot of noise. It was alert and interested in its surroundings and was playing with a red orchid it had grabbed from an adjacent bush.

The monkey was one of a series which had been given a fetal-brain graft with cells from the substantia nigra of a monkey fetus. Using stereotactic surgery, the material had been inserted directly into the striatum and, to everyone's delight, it had worked. Within a few months the monkey had started to improve, and the improvement appeared to be permanent.

Since 1985, Redmond had been responsible for a major neural-transplantation research program for Yale University using the MPTP primate model. For both practical and political reasons, the monkey work was done at a research station in St. Kitts, where there was an indigenous monkey population. Monkeys thrived in such a climate, and handling them was much easier than in the northeastern part of the United States. Also, it avoided having to deal with the animal rights fanatics. If this lab were in New Haven, Connecticut, there would probably be all kinds of demonstrations and hate mail to deal with.

Many animal rights advocates would prefer that this operation be carried out on humans. Paradoxically, they had less problem with the idea of experimenting on humans than on animals. But Redmond, like most biomedical researchers, knew that animal experiments were essential. Without them there would be no medical research. This was especially true in brain research. Almost everything that had been learned had been teased out of brilliantly designed animal experiments. And using fetal tissue to reverse parkinsonism was just the latest advance.

Redmond had sacrificed some monkeys and sent the brain tissue back to his collaborator John Sladek at the University of Rochester for analysis. Sladek had confirmed that the fetal grafts had taken. Slides of the brain showed clearly and unambiguously that the new fetal cells were not only surviving but prospering. They were sending out axons to surrounding host cells, where they made connections. Most remarkably, the cells seemed to be making the right connections. Tropic, or guidance, factors released by the trauma of the surgery seemed to be guiding the nerve fibers to make the appropriate connections.

Redmond realized where this work might lead. If this work panned out, many neurological diseases might one day be helped by fetal-tissue transplants. Apart from Parkinson's disease, there were Alzheimer's and Huntingtons', not to mention stroke victims and spinal cord injuries. Moreover, the U.S. population was aging fast. In twenty years, by the time Redmond was 60 to 65 years old, there might be two or three times as many patients with Alzheimer's disease to treat.

But the road ahead was going to be very difficult. To do fetal surgery on monkeys in St. Kitts was one thing. To do them on human patients in the United States, where there was a militant pro-life lobby as well as an animal rights movement, would be quite another.

San Jose, California, Spring 1987

In the five years since he had first seen George Carillo, Langston's group had established themselves as an internationally respected research team. They had published more than fifty scientific papers, including several review papers about the state of Parkinson's disease research. Their work had probed how good a model MPTP parkinsonism was of Parkinson's disease. They had studied the intellectual changes of MPTP cases and compared them with idiopathic Parkinson's patients. They had pursued chemical studies which confirmed work by Javitch and Snyder that MPP+ had to be taken up via the dopamine uptake system into dopaminergic neurons to cause toxicity. They did research on how MPTP and MPP+ were distributed in the nervous system.

They had investigated the relationship of MPTP parkinsonism and aging, demonstrating that the effects of MPTP were age related, like idiopathic Parkinson's disease. They showed that MPTP produced inclusion bodies in the neurons of older animals that were reminiscent of Lewy bodies, the cardinal pathological finding of idiopathic Parkinson's disease.

Several of the papers marked important advances. The PET scan study of asymptomatic MPTP patients carried out with Donald Calne was the first-ever report that dopamine deficiency could be visualized preclinically in human beings.

The most important potential therapeutic initiative to come out of MPTP, however, was deprenyl. Ever since the early research on MPP+ and deprenyl had been published, Parkinson's disease patients had been asking Langston if they could try deprenyl. These impassioned requests led Langston to set up a controlled trial to scientifically evaluate whether deprenyl could protect dopaminergic neurons, thus slowing the progression of Parkinson's. He managed to get funding from pharmaceutical companies in Europe and the United States to conduct a limited trial, and enlisted Jim Tetrud to help design and run it.

On June 9, 1987, shortly after they had begun recruiting patients for their study, researchers at the University of Rochester and Columbia announced they were planning a massive multicenter trial to evaluate the efficacy of deprenyl, to prove whether it really did slow the progress of the disease for patients in the earliest stages of Parkinson's disease. Called the DATATOP Study (Deprenyl and Tocopherol Antioxidative Therapy for Parkinsonism), it planned to recruit eight hundred patients and randomly assign them to one of four treatment groups: active tocopherol (a biologically active form of vitamin E which many people thought acted as a scavenger of free radicals); active deprenyl; active tocopherol and deprenyl combined; or a placebo. These patients would be followed for up to two years to see which group's Parkinson's disease progressed the slowest.

Langston's relationship with the Santa Clara Valley Medical Center had not improved since the debacle with Barbara Arons, the head of psychiatry. Despite increasing international recognition, VMC didn't regard him as anything special. In fact, they wouldn't even give Langston an office in the hospital. Ian Irwin ran the lab research at the Institute for Medical Research, and one day a week Langston held a Parkinson's disease clinic. He no longer had any affiliation with Stanford.

The last two years had been the most difficult of Langston's life. The poor accommodations were the least of his problems. Mostly, he worried about paying the bills. Except for his one day a week in clinic, all of his support had to come out of research grants. His group was therefore highly vulnerable. There was no safety net for the periods in between one grant and the next. Langston, now with three young children and a wife starting a residency in ophthalmology, was increasingly worried.

There were other problems. He knew that he was not giving adequate service to his patients. Supervising the research, writing grant applications, paying salaries, and seeing patients stretched him to the limit, and his clinic became booked a year in advance. He was faced with a choice: either close the clinic, or find a way of taking on additional staff and expanding it.

There was no practical way to expand the clinic through the Valley Medical Center, because the county had a hiring freeze at the time. The only option was to take the whole operation outside and open a fee-for-service clinic. Langston dreamed of a place where, under one roof, his group could carry out patient care and do basic research. A place where he would be free from the politics and petty bureaucracy of VMC and Stanford. A place where there would be a steady cash flow which guaranteed salaries and benefits, even if a grant proposal was rejected.

Silicon Valley is a place where people take dreams seriously. And to one of Langston's new friends, Jim Bottomley, this dream didn't seem ridiculous at all.

The highly successful owner of a Budweiser distributorship, Bottomley had become interested in Parkinson's disease when his father developed it. He had started to support research, including some of Langston's projects. As Langston came to know Bottomley, he realized that in addition to his extraordinary business acumen, he was one of those rare individuals who could be described as wise, who could truly see the forest and not just the trees. Bottomley's advice could be invaluable, so Langston decided to ask for some. He asked him for advice on how to achieve his dream of a comprehensive treatment and research center for Parkinson's disease.

"To do what you really want to do, you should start your own foundation." A foundation, Bottomley explained, could do all of the things necessary to survive, from raising funds for research from grants and corporations to offering a fee-for-service clinic.

Things happened very quickly. Bottomley helped to recruit a group of local business leaders for the board of directors, and found a suite of offices on Moorpark Avenue close to VMC that was leased by Southland Corporation (owner of 7-Eleven Stores), which they had vacated even though there was two more years left on the lease. Since it was for a nonprofit, Southland agreed to sublease the space for almost half price. It wasn't suitable for lab work, so Ian Irwin and the other basic researchers would stay at the Institute for Medical Research on the VMC campus for the time being, with the foundation's administrative and clinical work being done at the new offices on Moorpark.

In January 1988, a stack of papers arrived for Langston's signature, committing them to two years rent. The nonprofit had not yet been completed as a legal entity, and Langston didn't have a penny to cover the rent. As he sat there pen in hand, he began to panic. How could he take on this kind of responsibility?

At that moment, Bottomley walked into the room, saw Langston's face, and immediately grasped the situation. "Don't worry, Bill," he said with a twinkle in his eyes. "I'll stake you for the first year. After that, if it's a success, it will be your success. But if it's a flop, it will be your flop."

Langston took a deep breath and signed the papers.

After that, they went to work. Jim Bottomley and his foreman spent evenings measuring the newly leased space for the new clinic rooms. Langston's nurse bought pictures at K Mart to hang on the walls. They had signs made and stationery printed and phones put in. They hired administrative staff. Jim Tetrud joined Langston to run the clinic and they hired an occupational therapist to help with patient care.

As Langston walked in through the door of his California Parkinson's Foundation every morning, he felt scared but proud of what they had achieved. So much had come out of MPTP research, from the environmental hypothesis to radical new therapies like deprenyl. But in spite of the many positive things which had come out of the discovery that MPTP was neurotoxic, Langston knew that none of it had yet helped the individuals who had started it all—his patients with MPTP parkinsonism.

All six of them were in terrible shape. Langston had been to Vacaville prison to examine David and Bill Silvey and Toby Govea. It had been a depressing experience. The main problem was that their medication was not being well balanced. Balancing the L-dopa with bromocriptine, a drug that stimulates dopamine receptors directly, and other drugs is a very challenging task. The dosages have to be large enough to be effective but no larger than necessary to prevent the buildup of side effects. If it is not done carefully, the patient goes downhill fast. Bill Silvey was having profound problems with hallucinations, and Dave was having severe dyskinesias. Toby was becoming more and more paranoid, suffering from ideas that people in the jail wanted to kill him.

Connie had recently been hospitalized for a drug holiday, a period off medication. When a Parkinson's patient deliberately stops taking L-dopa, he or she freezes up, and is hospitalized in bed and given round-the-clock nursing care. The hope is that the patient can start on a lower dose after the drug holiday, thereby reducing the side effects. But in Connie's case, the drug holiday had not worked.

Langston reluctantly made a difficult decision. Connie's hallucinations were so severe, so disabling, and so violent and dangerous to her family that after the drug holiday, he was going to radically cut back her medication and eventually discontinue it altogether if necessary. He knew he was sentencing her to an indefinite imprisonment in her body, but it was the lesser of two evils, since neither Connie nor her family were willing to tolerate the side effects any longer.

Then there were George and Juanita. Juanita had dropped out of the scene more or less completely, still living in Montana with her family on the Rocky Boy reservation. Langston hadn't seen her for a couple of years.

His "best" patient was George. Since getting out of prison, George had been a model patient. Every month he came in for his checkup, and his medication was being very well handled. George had realized that his future was tied up with Langston's unit. But the truth was that George was in terrible shape. He could hardly move, and was constantly having threatening hallucinations, seeing snakes and fire every day. Langston wondered how anyone could tolerate such a condition.

Langston had absolutely no hope to offer any of the MPTP patients. For a brief time, he had wondered whether the much-talked-about adrenal transplants might offer a solution. Recently he had seen a 45-year-old woman, Mrs. Maria Moldonado, who had gone to Mexico City to have the operation done by the "master," Dr. Madrazo himself.

Tragically, this patient had lapsed into coma within hours of the operation, and had remained in a vegetative state for the next eight months until she died of pneumonia. During that period she made only piteous moaning sounds when stimulated.

Mrs. Moldonado's brain tissue was sent to Lysia Forno for postmortem analysis. The results were stunning. First, the graft was not in the place that Madrazo claimed to have inserted it. According to the medical charts, Madrazo had placed the graft on the right side of the patient's brain, in the caudate nucleus. But when Forno examined the brain, she found that instrument tracks had narrowly missed the target area, crossed the corpus callosum—a thick collection of white-matter tracts (nerve cell processes) that connect the two sides of the brain—then traveled into the other side of the brain, where there was a large area of infarction (stroke) in a structure called the thalamus. After days of searching, Forno finally found what appeared to be the remains of the graft—not in the caudate, but in the fibers that connected the two sides of the brain.

No, based on the current evidence, Langston would not recommend any of his MPTP patients for an adrenal graft. The benefits, if any, were temporary, and the risks considerable. The prospects for the MPTP patients, especially Connie—who Langston thought might not survive much longer given her frail state of health—looked grim.

17

Doing the Unthinkable

Brussels, Belgium, October 27, 1986

It happened quite by accident. Langston had been invited to a conference in Belgium on neurodegenerative diseases of aging organized by the famous Houston surgeon Michael E. DeBakey. As part of the program, the scientists had been invited to a state dinner at the Summer Palace hosted by Princess Liliane. At six o'clock a bus arrived at the hotel and a group of eighteen distinguished scientists in tuxedos boarded it.

As the bus left the city, it began raining very heavily. Langston looked out of the window and thought of California. Seeing other countries had convinced him that he lived in the most wonderful place in the world. Where else has such beautiful weather and spectacular scenery? Where else but California could you go surfing and skiing in the same day? On the other hand, where else but California might you expect someone to make designer drugs? On trips like this, Langston often thought about the addicts, especially Connie Sainz. He had grown fond of the Sainz family and felt bad that there was nothing he could do. When he saw Connie confined to a chair in front of the television, unable to move or talk, it nearly brought tears to his eyes. He sometimes felt guilty that her tragedy had brought him such good fortune. Here he was in Brussels being honored largely because of the tragedy that had befallen Connie and the others in 1982.

Langston enjoyed traveling to exotic parts of the world and meeting other scientists, but greatly regretted the time taken away from his family. In his first marriage he had spent too much time working, time that should have been spent watching his three daughters grow up. Now he wanted to be careful that he didn't miss seeing his three sons, James (age 5), Mikey (age 4), and Danny (age 2), do the same. For this reason he was doing this trip to Brussels in forty-eight hours flat.

Balancing home and a career had been especially difficult for Langston over the past three years. Starting the California Parkinson's Foundation had been like starting a business. He wasn't just a doctor/researcher, he was an entrepreneur. He constantly worried about failing. It had given him a respect for the businessmen who started up high-tech companies. But it was also very satisfying. The foundation was not only a growing clinical center for patients with Parkinson's disease and other movement disorders, it was making a significant contribution to basic research. Currently they were doing basic research into the metabolism of MPTP (with the hope of developing therapeutic drugs), work in Parkinson's disease and aging, and research into the epidemiology of Parkinson's disease.

The rain continued to beat against the window. It was now an hour since they had left the hotel, and Langston began to suspect that the bus driver might be lost. One of the scientists on the bus politely asked the driver, "Do you know where we're going?" The driver assured them that he knew the way to the Summer Palace.

Half an hour later, it was clear that he didn't. Now late for their dinner, the scientists began to get anxious. Looking out of the window, Langston sensed that the bus was traveling on smaller and smaller roads. It seemed impossible that this was the way to the Summer Palace. Eventually, the driver turned down a narrow lane that looked like a cow path and was forced to admit that he was well and truly lost.

The driver apologized for his mistake and looked for a place to turn the bus around. By now it had been raining heavily for over an hour and a half, and the soil was soft and sticky. The heavy bus reversed into a field, sank into the mud, and became completely stuck. They were stranded in a field in Belgium.

The eighteen scientists in tuxedos, now an hour and a half late for their state dinner, sat in the dark listening to the rain pounding down on the tin roof of the bus, with no idea what to do. When the driver asked if any of the scientists would mind getting out to push the bus—an idea that was firmly rejected—Langston began to feel as if he were in a Buñuel film in which the situation starts out reasonably normal and then gets more and more surreal.

Eventually the driver said that he would go and try to get help, and off he went carrying his coat over his head, leaving them sitting in the dark. Langston got up to stretch, walked up and down the aisle a few times, and then sat down next to someone he didn't know and introduced himself. "I'm Bill Langston."

"Nice to meet you," came the reply, the accent clearly Scandinavian. "I'm Anders Björklund."

The pair knew a little of each other's work, but had never met. They began talking about their research and soon became totally involved in the conversation. Björklund knew about the MPTP cases and enjoyed listening to Langston tell their story. For his part, Langston found the work Björklund was doing fascinating.

Listening to Björklund describe the program at Lund, Langston realized that they had built something very unusual—a research program which seamlessly combined first-class scientific and clinical research; a place where

Björklund's lab work informed Olle Lindvall's clinical decisions, and vice versa.

Björklund expressed his deep reservations about the Mexican work and adrenal transplants in general. He went on to talk about fetal-tissue transplants and their discussions about doing human cases.

"There are so many things that we don't know," Björklund remarked with a hint of despair. "Apart from the monkey MPTP experiments, and our rodent work, we have so little to go on. We don't know exactly how much material to use, we don't know exactly where to put it. We know we want to put it in the caudate or putamen, but exactly how to spread it out isn't clear. We are not sure about the risks of immunological rejection. One thing we are pretty sure about is the age of the fetal material. It must be six to eight weeks after fertilization, otherwise we don't think there is much hope."

Langston could barely make out Björklund's shadowy profile in the dark. "Are you worried that the continuing Parkinson's disease might be attacking the graft?"

"Certainly. In the MPTP monkeys—or in your MPTP patients, for that matter—the cause of their parkinsonism is no longer there. The damage is done. The disease is over. But in ordinary Parkinson's disease, there is the definite possibility that whatever caused their Parkinson's disease could still be present and the disease may continue to progress. So this might mask any effect of the graft or it might even attack the graft itself. We don't know."

Deep in Langston's mind something clicked. Perhaps, he thought, one day there might be hope for Connie and the other MPTP patients after all.

Neurology Library, Lund Hospital, December 1988

Olle Lindvall looked at his colleagues sitting round the table: Anders Björklund, Patrik Brundin, Håkan Widner, and Stig Rehncrona, the core of the Lund fetal-tissue transplant task force. He was proud to have been associated with such a group. In the three years since they had formed the task force, they had made enormous progress.

The political hurdles had been crossed quite easily. They had asked the Swedish Society of Medicine to look into the ethical issues of fetal-tissue transplant and to determine whether such transplants would be acceptable to Swedish society. Abortions had been legal in Sweden since 1975, although there were more restrictions than in the United States. While first-trimester abortions (up to twelve menstrual weeks) were available without restriction, pregnancies of twelve to eighteen weeks required a special investigation (usually a formality) before abortion was sanctioned. From eighteen to twenty-two weeks, only medically indicated abortions were allowed—for example, where inherited disorders were involved.

For this reason, the vast majority of the thirty-two thousand abortions performed annually in Sweden were first-trimester terminations.

The Society of Medicine's ethical committee, consisting of lay people, politicians, philosophers, lawyers, scientists, clinicians, and religious interests, had begun their deliberations in 1985. Opinions were invited from individuals and groups who felt strongly about the subject, but remarkably, it had not generated much controversy or media coverage. Within a year, the committee had come up with preliminary ethical guidelines.

The guidelines were meant to apply to fetal tissue in general, not just brain tissue transplants. The committee essentially tried to separate the issue of abortion from the issue of how to use fetal tissue. It was decided that there should be no connection between the donor and recipient. A daughter would not be allowed to direct her aborted fetus to her father suffering from Parkinson's disease or Alzheimer's disease, for example. A woman would not be allowed to sell her fetus for money. The transplant must not influence the abortion in any way. Only after a woman had decided to have an abortion might her consent be sought to use the material for a transplant.

It was agreed that only work on dead fetal tissue would be permitted. A fetus could not be kept alive in utero for the purpose (or convenience) of scientific research. Abortions would be carried out in the routine way—in Sweden, the suction method is the most common method for first-trimester abortions.

There had been little negative reaction. Only a very few anti-abortionists in Sweden had argued that fetal-tissue transplants would make abortions more likely. Indeed, some who were against abortion on religious grounds were nevertheless in favor of using the tissue to help someone rather than throwing it away. Olle Lindvall's father, then in his eighties, was a priest in the Swedish church. Father and son had discussed the issue at length together and had reached agreement that it was ethical to go ahead.

Meanwhile, the Lund task force had fully explored some of the technical hurdles and had carried out thousands of experiments on rats. Dr. Håkan Widner, a young physician with a doctorate in immunology, had been recruited for his special knowledge of brain immunology. This was one of the crucial areas, yet the action of the immune system within the brain was still not fully understood. Most tissues of the body are strongly immunogenic. For example, if skin is transplanted from a black mouse to a white one, it is rejected. But the same thing doesn't seem to happen with brain tissue: a transplant of brain tissue from a black mouse to a white mouse will survive. Because of such experiments, it was commonly believed that the brain was an immunologically privileged site, one of several areas in the body where the immune system is relatively inactive.

It made sense. Large inflammatory responses in a delicate organ like the brain could have terrible side effects. The blood-brain barrier keeps out most of the cells from the immune system that recognize foreign cells—macrophages, for instance. But a number of animal experiments had been carried out which proved that the brain's immunological privileges were limited. Tissue rejection in the brain would occur if the immune system was challenged strongly enough. For example, if an animal was transplanted with a piece of skin from the back of another animal, and then some time later was transplanted again with nerve cells in the brain, the new nerve cells would be rejected. It all depended on how much the immune system within the animal had been stimulated. Until they knew more, it made sense to give immunosuppressant drugs to patients after the surgery, just to be on the safe side.

Patrik Brundin was in charge of the dissection and preparation of the fetal tissue. He had done hundreds of experiments with rat embryos and had become very expert at the procedure. Together with Anders Björklund and other colleagues, he had ascertained the optimal age of rat fetal tissue for transplantation. They had then gone on to graft human fetal cells into rats to calculate the best time frame for human fetal tissue. There was now very good experimental data which showed that the optimal time to transplant human fetal-brain tissue was from six to eight weeks following fertilization. This was critical. Go beyond that period, to ten weeks after fertilization, and very few cells survive. Go even further, to twelve weeks, and there is no survival at all. Abortions in Sweden tended to be carried out around the six- to eight-week period, so this was not a big problem.

A larger challenge had been making sure that the tissue was not infected. As they were implanting foreign tissue into a patient's brain, it was crucial that they didn't inject bacteria or viruses. The Swedes knew they had to ensure that the women undergoing abortion were not infected by any virus and that, once the abortion was carried out, the tissue was handled in a completely sterile way.

Then there was the actual dissection. This was no easy task. In nine cases out of ten, suction abortions disrupted the fetuses. Working under a microscope, Patrik Brundin had to first reassemble the central nervous system so that he could locate the target brain structures.

The entire fetus at eight weeks was only the size of a fingernail; the substantia nigra, smaller than the head of a pin. Brundin's task was to dissect only those cells whose destiny was to make dopamine. Very closely attached to the nigral tissue were cells which had a different destiny—to become cartilage, bone, skin. Brundin knew from animal experiments that if one isn't careful and grafts this tissue into the brain along with the substantia nigra cells, they grow into big bits of skin and cartilage. So it was critical that these fragments of so-called mesenchymal tissue were completely removed.

They had done some calculations based on animal experiments to estimate what proportion of these fragile fetal cells survived. The results were sobering. Even with all the precautions, only about 10 percent of the cells were likely to survive and function in the patient's brain. Some cells were lost at the time of abortion; some were lost during the dissection, preparation, and insertion; and some were lost during the period just after transplantation, when the cells don't yet have a fully developed blood supply. Because of this loss, the task force had decided that they would need to implant cells from several fetuses. Otherwise there might be no effect at all.

Brundin and Björklund had perfected the techniques of tissue dissection and processing through work on thousands of rat embryos—which are similar in size to human embryos. Because the surgery was difficult, they carried out thirty full rehearsals before attempting a human transplant so that everyone, from gynecology to medical imaging, knew what they had to do.

A few months after the 1987 Rochester conference, while surgeons all over the world were rushing to do adrenal transplants, the Swedes quietly carried out the first human fetal-tissue transplants. The first operation was performed in the second week of November 1987, on a 47-year-old patient, followed a month later by the second operation, on a 54-year-old. Both patients were women who had suffered from Parkinson's disease for fourteen years, having contracted the condition while young (one being diagnosed at age 33, the other at 40 years of age). The neurosurgeon was not Erik-Olof Backlund, who had carried out the adrenal transplants, but Stig Rehncrona from Lund. Backlund had originally agreed to serve on the Lund task force to help in the planning. But in 1986 he had moved to Bergen, Norway, and decided, for practical reasons as well as ethical objections, to withdraw from the team.

The Swedes had cut no corners. The purpose of these operations was not only to help the patients but to extract as much scientific information as possible. The two patients had been carefully selected for a number of reasons.

Both had advanced disease with significant on/off fluctuations. Even with medication, the 47-year-old could move well only for a few hours a day. The 54-year-old had been forced to give up her job as a computer clerk in 1982 because she could not move quickly enough. By 1987, she was in the off state for about half the day.

Both women experienced a lot of rigidity and bradykinesia with relatively little tremor, and these symptoms affected one side of the body more than the other. This one-sidedness was important to the team. Because the Parkinson's of the 47-year-old, for example, was predominantly on the right side of her body, a transplant to the left side of her brain (which controlled the right side of her body) should, if successful, have a discernible clinical effect.

Both patients were relatively healthy. Apart from their advanced Parkinson's disease, they had no medical complications or psychiatric problems. They were good candidates for surgery.

After they had been selected, the patients had been stabilized on a level of L-dopa medication and subjected to a battory of quantitative tests—walking up and down, hand flipping, etc.—at regular intervals for six months prior to surgery. They had been flown to London's Hammersmith Hospital for a baseline PET scan. Radioactive fluorodopa had been injected into their veins and the rate at which it was absorbed in the striatum had been visualized and measured. The pictures showed the extent of the dopamine loss in the striatum of both patients. In patient number two, for example, whose symptoms were worse on her left side, especially her left arm, the lowest amounts of fluorodopa were found in her right putamen. This data was kept so that it could be compared with the patients' PET scans a year after surgery.

In the operation, each patient received material dissected from four fetuses. The fetal material was mixed with a chemical called trypsin, dissociating the cells into a liquid suspension which was drawn up into a cannula by the neurosurgeon and implanted into one side of the brain. The cells were not implanted into the substantia nigra region, for two main reasons. First, it is a very difficult and dangerous site to reach, being a part of the brain stem. Second, there was good reason to believe that grafting to this region would be ineffective. Dozens of experiments in rats had shown that cells grafted to the substantia nigra did not reverse their Parkinson's-like symptoms, because the grafted nerve fibers could not grow such distances to reach their targets in the striatum. In contrast, grafts placed in the striatum itself worked spectacularly well in rats.

The striatal site with the lowest fluorodopa uptake in the PET scans (and therefore the greatest dopamine depletion), was the putamen. Additionally, the team felt that it was important to re-innervate the caudate region. After long deliberations, they had decided to make three implants for each patient, shared between these two sites—two into the putamen and one into the caudate.

The operations were carried out discreetly, with no publicity—that would come only after publication of the results. Then the team would methodically follow the patients for a year, periodically testing them using standard rating scales.

Despite continual phone calls from the press, the Lund team had refused all comment for over a year while they patiently gathered data. Now the time had come to weigh the results for presentation at a scientific meeting in Jerusalem in early June of 1989. Today, in the small library of the neurology department, they were reviewing the data from these first two cases and planning their statement. They knew that there would be a lot of interest in what they had to say. With adrenal transplants discredited, this was now the only hope.

The results were very modest. There had been some improvement in the two patients. The first patient, the 47-year-old, had experienced severe off periods prior to the operation, when she was completely unable to walk. Now, one year after the operation, although she had the same number of off periods, she was able to walk during virtually all of them. She also was less rigid and could move a little more quickly. On the other hand, measurements of dopamine absorption in her brain via the PET scan method showed little change.

The second patient could also walk a little better, as measured by the clinical tests, but the effects were small and varied over time. Her PET scan showed almost no change.

The improvements, a year after surgery, were encouraging compared to the adrenal transplant operations, but disappointing compared to the results reported in animals.

The members of the task force had spent hours discussing the results. Any number of factors might be involved. Had they used enough fetal tissue? Had enough of it survived to make a difference? Had they placed it in the right location? Should they have spread out the graft more so that it made a wider set of connections with surrounding tissue? Should they have put grafts into both sides of the brain—was it reasonable to expect a bilateral effect from a unilateral graft? Had they allowed enough time for the immature cells to grow?

And there were other, more worrying possibilities. The rats and monkeys, in whom fetal grafts worked, were young and otherwise healthy. The cause of their parkinsonism—toxins like MPTP—was known. The patients with regular Parkinson's disease were middle-aged or elderly. The cause of Parkinson's disease in the two patients operated on was unknown, and since their disease was progressive it was quite possible, even likely, that the cause remained in their bodies. If this was the case, there were at least two consequences: either the graft was making a small difference but its beneficial effect was fighting a general decline, or the graft was being attacked by whatever caused the Parkinson's disease in the first place. Until they resolved some of these questions, they couldn't justify many more such operations.

18

Embryonic Conflicts

When a human sperm fertilizes a human egg, a remarkable process is set in motion. Twenty-three chromosomes from the father and twenty-three chromosomes from the mother combine in the nucleus of the first cell. These chromosomes contain the genetic information to make a new person. The egg divides into two parts, then four, then eight. After about seven days the embryo becomes a ball of hundreds of cells called the blastula. Up to this stage each cell in the ball has the potential to develop into any of the many tissues that make up the human body—be it bone, heart, blood, kidney, or brain. It is only at about two weeks that the cells lose this general potential and take on a particular fate.

The embryo starts to fold inward in a process called gastrulation. In this critical process, which the British embryologist Lewis Wolpert has called "the most important stage of life, far more significant than birth, marriage or death," cells divide up into three groups—the endoderm, which produces the gut and lungs; the mesoderm, which produces blood, connective tissue, and kidneys; and the ectoderm, which makes the brain and spinal cord.

The first ectodermal structure to be built is the neural plate, a sheet of some 125,000 cells, which elongates, forming a groove which soon becomes a tube called the neural tube. This tube develops into the spinal cord. Meanwhile, three lumps appear at the head end of the embryo—the first parts of the brain. All this has happened by four weeks after fertilization.

Now the cells in the brain and spinal cord begin to specialize. They differentiate into different kinds of nerve cells, including the dopamine-making cells of the substantia nigra.

To help the nerve cells find their right place in the brain and spinal cord, a special group of cells called radial glia guide them. The radial glia act like temporary scaffolding. Once in position, the nerve cells sprout dendrites (short processes which receive impulses) and axons (longer processes which extend from a cell body to target cells the neuron is to communicate with). The axons grow and find their destinations with the help of chemicals—so-called neurotrophic (nourishing) and neurotropic (guiding) factors.

By this time—about eight weeks after fertilization—the number of nerve cells, or neurons, in the midbrain is essentially fixed; no new neurons will be produced for the rest of a person's life. In fact, more than half of these cells will die off by the time the fetus reaches term. During the remainder of pregnancy the growing neurons must compete with one another for a limited number of available synaptic sites, places where the nerve cells hook up to communicate with other nerve cells. Through this process, the number of neurons matches the needs of the developing nervous system.

When a baby is born, in the order of 100 billion neurons remain. The neurons cannot increase their numbers by division. These brain cells are the only ones the individual will have, and if they die, no new ones will be made to replace them.

Anders Björklund's careful experiments had shown that the only nerve cells which can possibly be used to repair a parkinsonian brain are fetal cells, about six to eight weeks old. Repeated experiments grafting cells older than this invariably ended in failure—the cells didn't survive. In fetuses younger than six weeks, it was virtually impossible to locate the substantia nigra cells. Six to eight weeks following fertilization seemed to be the optimal time. Only in a fetus of this age could one find cells that are differentiated and know their fate (for example, to make dopamine) but that can survive, grow, and make connections with surrounding neurons.

National Institute of Neurological and Communicative Disorders and Stroke, Bethesda, Maryland, October 1987

In the fall of 1987, Ed Oldfield, a scientist at the National Institute of Neurological and Communicative Disorders and Stroke (NINCD), had sought permission from his boss, Dr. Irwin Kopin, director of intramural research at NINCD, to transplant human fetal tissue into the brains of patients with Parkinson's disease.

While Oldfield lined up the first group of patients, Kopin sought and obtained the approval of the institute's human subjects review board in the usual way. But given the explosive nature of anything related to abortion, Kopin decided it would be politically prudent to inform his boss, James Wyngaarden, head of the NIH, of their intentions.

Scientists in the United States had been using fetal tissue for most of the twentieth century, with little or no outcry. Doctors had transplanted fetal thymus for DiGeorge's syndrome (a rare and usually fatal inherited disease involving multiple abnormalities of organs and glands) at the turn of the century. Fetal tissue was used to prepare polio and rubella vaccines in the 1950s, and more recently to make monoclonal antibodies. Scientists interested in understanding fetal development and genetic anomalies needed to work with human fetal tissue. There was no other way to do this kind of research. Fetal tissue was also being used to culture cell lines which were widely utilized for everything from toxicity testing and manufacturing pharmaceutical products to AIDS research.

The NIH was a big player in this research. At the time of Oldfield's request, NIH spent some eleven million dollars a year spread across ten of the national institutes on 116 different grants and contracts involving the use of fetal tissue.

But the idea of using aborted fetuses to save lives (in what might one day become a routine clinical procedure) had huge implications, and Kopin wanted to know if his boss supported the experiment.

Now the hot potato was Wyngaarden's responsibility. Wyngaarden was fully empowered to okay the research, or could if he chose refer the matter upward. Wyngaarden took the politically prudent course and referred Kopin's request to his superior, Assistant Undersecretary of Health Robert E. Windom, thereby bringing the latest neuroscience research to the attention of the Reagan administration.

Nothing happened for a few weeks. Then, on March 22, 1988, Windom sent a carefully worded memo to Wyngaarden. "This proposal raises a number of questions—primarily ethical and legal—that have not been satisfactorily addressed, either within the Public Health Service or within society at large . . . I am withholding my approval of the proposed experiment, and future experiments, in which there is performed transplantation of human tissue from induced (elective) abortions." From that day on, no NIH-funded research would be allowed involving human fetal-tissue transplants—not just in Parkinson's disease, but in any disease. Windom then ordered Wyngaarden to convene a panel of experts to examine the legal, ethical, and medical aspects of fetal tissue research in particular, addressing ten questions.

1. Is an induced abortion of moral relevance to the decision to use human fetal tissue for research? Would the answer to this question provide any insight on whether and how this research should proceed?

2. Does the use of fetal tissue in research encourage women to have an abortion they might otherwise not undertake? If so, are there ways to minimize such encouragement?
3. As a legal matter, does the very process of obtaining informed consent from the pregnant woman constitute a prohibited "inducement" to terminate the pregnancy for the purposes of the research—thus precluding research of this sort, under HHS regulations?
4. Is maternal consent a sufficient condition for the use of the tissue or should additional consent be obtained? If so, what should be the substance and who should be the source(s) of the consent, and what procedures should be implemented to obtain it?
5. Should there be and could there be a prohibition on the donation of fetal tissue between family members, or friends and acquaintances? Would a prohibition on donation between family members jeopardize the likelihood of clinical success? (In other words, what if any immunological advantages are there for a transplant within a family?)
6. If transplantation using fetal tissue from induced abortions becomes more common, what impact is likely to occur on activities and procedures employed by abortion clinics? In particular, is the optimal or safest way to perform an abortion likely to be in conflict with preservation of the fetal tissue? Is there not any way to ensure that induced abortions are not intentionally delayed in order to have a second trimester fetus for research and transplantation?
7. What actual steps are involved in procuring the tissue from the source to the researcher? Are there any payments involved? What types of payments in this situation, if any, would fall inside or outside the scope of the Hyde Amendment?[1]

[1]The Hyde Amendment, introduced by Rep. Henry, R-Ill, prohibits the federal funding of abortions except when necessary to save the life of the mother.

8. According to HHS regulations, research on dead fetuses must be conducted in compliance with State and local laws. A few States' enacted versions of the Uniform Anatomical Gift Act contain restrictions on the research applications of dead fetal tissue after an induced abortion. In those States, do these restrictions apply to therapeutic transplantation of dead fetal tissue after induced abortion? If so, what are the consequences for NIH-funded researchers in those States?
9. For those diseases for which transplantation using fetal tissue has been proposed,[2] have enough animal studies been performed to justify proceeding to human transplants? Because induced abortions during the first trimester are less risky to the woman, have there been enough animal studies for each of those diseases to justify reliance on the equivalent of the second trimester human fetus?
10. What is the likelihood that transplantation using fetal cell cultures will be successful? Will this obviate the need for fresh fetal tissue? In what time frame might this occur?

They were good questions and they needed to be debated thoroughly. Most scientists, Irwin Kopin included, thought the panel a good idea. It made perfect sense to fully examine new areas in advance and think through the implications of a new therapy. This process was under way in a number of other countries, including Sweden, the United Kingdom, and Australia. A temporary moratorium on NIH funding for projects like Ed Oldfield's for a few months was not unreasonable. As instructed, Wyngaarden set up a panel of experts, under the chairmanship of a retired federal judge, Arlin M. Adams, and hearings were set to begin in September 1988.

Most interested scientists thought this was just a temporary holdup in their work. It seemed unlikely that the panel

[2]Parkinson's disease and juvenile diabetes.

would ban the use of this material. Medical researchers trying to find ways of treating Alzheimer's, Huntington's juvenile diabetes, epilepsy, stroke, muscular dystrophy, spinal cord injury, pancreatitis, immunological diseases, AIDS, and of course Parkinson's disease depended on access to human fetal material. It would make little sense to interfere with such work.

But early in September 1988, just days before the panel held its first meeting, something happened which indicated that the road ahead would not be easy. The press got hold of a document written by Garry Bauer, assistant to President Reagan for policy development. The document was a draft of an order forbidding all federally funded research on tissue from aborted fetuses. The NIH panel had yet to start its deliberations. What was a confidant of the president doing writing an order banning federally funded fetal-tissue research before the panel had considered the subject and reported its conclusions? Did this mean that whatever the panel decided, the administration would ban this research? Did it mean the temporary moratorium might not be so temporary after all?

While presidential spokesman Marlin Fitzwater later claimed that the draft order was only a "first cut" and that "it was unlikely that the President will issue a no fetal research order until hearing from the human fetal tissue research panel," scientists were not reassured. Fully informed about the Swedish transplants in late 1987, several U.S. medical teams had been on the verge of doing human fetal-tissue transplants. In addition to Ed Oldfield at NIH, Eugene Redmond's team at Yale were anxious to see if their primate work could be transferred to humans. Also keen to start fetal transplants for Parkinson's disease was Curt Freed at the University of Colorado. Redmond and Freed were somewhat unnerved by the moratorium. They were faced with an uncertain future. If the administration was against such research, then no matter what the panel decided there would be no way to do such operations in the United States except with private money. The operations were not cheap. Rough estimates put the figure somewhere between thirty and fifty thousand dollars.

The politics of this situation were volatile. On one side were political conservatives and the right-to-life lobby. On the other side were the pro-choice lobby and a huge number of patient advocacy groups, including Parkinson's disease, AIDS, juvenile diabetes, epilepsy, stroke, muscular dystrophy, Alzheimer's, and immunological disorders. Both sides felt they had a lot to lose.

California Parkinson's Foundation, San Jose, California, January 1988

Ever since meeting Björklund on the bus in Belgium, Bill Langston had wondered whether fetal material might offer a chance for his MPTP patients. He was especially worried about Connie Sainz. The last time he had been down to Greenfield, she had looked terrible. He was increasingly worried that she wasn't going to live much longer.

The Swedes had seemed very interested in the MPTP cases. They not only were moved by their story, but also realized that these patients held enormous scientific interest. They might hold the key to why the first human fetal-tissue transplants had been so disappointing. The MPTP cases were young and the cause of their parkinsonism—MPTP—was known. Unlike ordinary Parkinson's disease cases, they had no continuing disease which might attack the fetal graft. If fetal transplants could ever work in humans, then they really should work in the MPTP cases.

The process would be difficult. First the Swedes would have to come to California and evaluate all of the patients, three of whom were residing in Vacaville prison. Then, provided they thought that any of the patients were suitable for a scientific trial, they would begin their preparations. Langston would need to get the protocol approved by the local human subjects committee. A similar review process would take place in Sweden. Then there would be problems financing the procedures—a costly airfare to Sweden, the operations, the postoperative care, and all the follow-up. Then there were the problems of actually taking very sick patients on a fifteen-hour journey. And of course there was the controversial matter of the fetal tissue. As if he didn't have enough problems, there was the NIH moratorium and the ensuing controversy, which looked like it was going to be brutal. This might be enough to stop the operations, even though they were to be done in Sweden.

Meeting of the Human Fetal-Tissue Transplantation Research Panel Consultants to the Advisory Committee to the Director, NIH, September 14, 1988

Retired Judge Arlin M. Adams sat listening to Professor Lars Olson from the Karolinska Institute in Sweden. Olson had come over to brief his panel about the latest research going on in Sweden and elsewhere. After reviewing the scientific research that he and his colleagues at Lund had carried out over two decades, Olson told them about the latest fetal-tissue transplant operations done in Lund. His presentation was extremely clear and to the point. He then addressed Robert Windom's questions. Did he think that fetal-tissue transplants would encourage more women to have abortions?

"No. In Sweden approximately every fourth pregnancy is terminated by abortion. The use of cell and tissue grafting from aborted fetuses . . . would in no way cause a 'demand' for more aborted fetuses. Moreover it would be of no immunological advantage to graft tissue from a 'related' fetus . . . we still do not know whether immunosuppression is necessary at all."

What prospects were there for using cell cultures as a source and when might such cell lines be available?

"Cell cultures might eventually be successful; however, it is also possible that they will never be useful. One severe problem with cell cultures is that cells that can be maintained in culture, i.e., cells that have been 'immortalized,' are difficult to control after transplantation and can continue to grow, forming tumors. Thus, probably for many years to come, the only alternative for parkinsonian patients who have already lost their dopamine nerve cells is a grafting procedure involving fresh tissues."

Judge Adams thanked Professor Olson for his illuminating presentation. He looked around the table at the assembled panel of scholars. In addition to ethicists, obstetricians, and scientists, there was a rabbi, a Roman Catholic priest, and an Episcopal clergyman. They were a very impressive group. In addition to Professor Olson, the panel would hear some fifty invited speakers, including sixteen representatives of public interest groups that had asked to testify.

Adams was a long-standing opponent of abortion, except in very limited situations in which the life of the mother was in danger. Yet abortion had been legal in the United States since the historic Supreme Court ruling in the case of *Roe v. Wade* on January 22, 1973. The court had ruled that a woman's right to choose to have an abortion was covered by the Fourteenth Amendment. Adams had serious reservations regarding the Supreme Court opinion, both in its reasoning and in its ultimate result, but the law was the law. The issue his panel had to decide was not whether abortion was ethically justified, but whether it was ethical and desirable to use the tissue generated in elective abortions for medical research.

Over the next three days the panel would receive a graduate education in fetal-tissue research. They would learn that approximately four thousand abortions take place in the United States every day—about one and a half million abortions a year. Ninety percent of these abortions—which take about ten minutes—are carried out in the first trimester of pregnancy, and are thus ideal for neural grafts, fetal-liver transplants, and fetal-thymus transplants. Yet in 1988, most fetal material was incinerated.

Whatever one's position on abortion, Judge Adams had to admit that on the surface this certainly seemed wasteful. But there were other issues the panel had to consider.

Under what circumstances might donation of fetal tissue be allowed, and would the practice of using fetal tissue to treat diseases like Parkinson's disease lead to an increase in elective abortion?

This was a very emotional subject. Much depended on how one procured the fetal tissue. In the past there had been some appalling experiments in which live fetuses had been aborted through hysterectomy. In one study, the fetuses had been decapitated and the heads kept alive artificially and studied. This joint U.S./Finnish project, which aimed to measure fetal metabolism, had been partly funded by the NIH. Another project had tried to keep a group of aborted fetuses alive in saline solution. The outcry about these experiments had led to a ban on research on live fetuses.

Nobody was suggesting that live fetuses be used in these operations. The argument most scientists used went as follows: A woman goes in for an abortion. The embryo is removed using the suction technique. The nurse collects the dead remains in a dish—some two ounces of tissue. No law has been broken. The question is what to do with the dead remains. The dead fetus is now a cadaver. It can be incinerated or, under the procedures laid down in the Uniform Anatomical Gift Act, it can be donated for transplantation or medical research. The mother, as next of kin, can sign a waiver and the fetal material can be used to save a life.

But not everybody accepted this argument. The interest groups like Right to Life making presentations before the panel argued that any mother capable of having an abortion had yielded her rights to her fetus. Moreover, if scientists really needed to use fetal tissue for their research, they didn't have to use elective abortions; they could use fetal tissue from spontaneous abortions, miscarriages, and ectopic pregnancies. Because these accidents of nature were not premeditated, it was perfectly all right to use the tissue to help another human being. What they objected to was using fetal tissue from an elective abortion, because this in some way validated abortion and, in their view, made it more likely.

Scientists appearing before the panel totally rejected the use of other fetal material, arguing that it was not of sufficient quality. While it is true that some seven hundred thousand first-trimester spontaneous abortions occur each year, in approximately 60 percent of these the fetuses have chromosomal abnormalities and many have been caused by infections. Other potential sources, such as ectopic pregnancies and stillbirths, rarely produce recognizable fetal tissue, or if tissue is recovered it is usually too mature (and thus unsuitable for graft survival). Implanting such unviable tissue into the brain or liver of a human being would, most scientists contended, be very dangerous and grossly irresponsible.

The majority scientific opinion held that first-trimester elective abortions provided the perfect source for most of the medical applications. The best age for fetal neural tissue for Parkinson's disease was six to eight weeks after fertilization, for fetal thymus it was eleven to fourteen weeks, and fetal liver needed to be between eight and fourteen weeks old. Only the fetal pancreas needed to be ten to twenty weeks, putting it well into the second trimester and making it somewhat more controversial.

Vacaville Prison, California, September 1988

The collaboration between Bill Langston and the Swedish team was going well. First Olle Lindvall and Anders Björklund had come out to California to meet with Langston and see some of the patients. Then Håkan Widner had been dispatched to do a thorough evaluation of all six MPTP index cases to assist with final selection. His first stop was California's main medical correctional facility, at Vacaville in northern California, where the two Silvey brothers and Toby Govea were all serving time for robbery.

Throughout the day, Widner carried out a full clinical

evaluation of the three young men and interviewed them in detail. They had heard about the operation and were excited about his visit, as this was the first hopeful piece of news they had heard from the world of medicine since the original tragedy.

All were in bad shape. The worst off was Bill Silvey. He regularly had terrible hallucinations and looked very weak. Like many prison patients, Bill Silvey was severely overmedicated, and if he remained in jail he could never qualify for a project like this. Even though David Silvey wanted to go himself, he lobbied hard for his brother during his evaluation session. "Please take my brother Bill. Talk to his parole officer; perhaps something can be worked out."

For a brief time Widner entertained the idea of actually talking to Bill Silvey's parole officer, but he soon gave up the idea. David Silvey was in somewhat better shape than Bill, but Widner was concerned that both he and his brother were still abusing drugs. The prison staff were convinced that they deliberately faked freezing spells to get more medication which they either abused or sold. At any rate, Bill was on enormous doses (4400 mg) of L-dopa. At peak concentration of the medication he hallucinated, had severe dyskinesias, and appeared deranged. He liked to bounce a ball off the wall, and at such moments he would chase his ball around the room manically.

Next Widner saw Toby Govea. Toby was very suspicious about the interview, and at first did not want to go through with it. He suffered from paranoid delusions; in his case it seemed likely that this was a sign of overmedication. In his present state it was doubtful that he could carry out the requirements of a complex trial.

A lot was riding on these proposed operations. They were experimental procedures, designed not only to help the MPTP patients but also to acquire knowledge which might end up helping millions of patients throughout the world. The team needed subjects who were reasonably reliable, who could be located at regular intervals, who would keep to an agreed dosage of L-dopa for at least six months before surgery and a year and a half after surgery, and who would avoid illegal drugs. A patient who disappeared halfway through the trial and began abusing drugs would invalidate the entire experiment. And this would be a disaster. The cost of the operation and associated care would add up to at least fifty thousand dollars per patient.

Widner doubted whether David, Bill, or Toby was capable of the required self-discipline. Furthermore, the odds of getting them out of Vacaville did not seem high.

He would not recommend any of them for the project.

Second Meeting of the Human Fetal-Tissue Transplantation Research Panel Consultants to the Advisory Committee to the Director, NIH, October 20–21, 1988

Judge Adams wasn't sure whether he would be able to get this panel to produce a consensus report. The question of fetal-tissue research had stirred up very strong passions. Throughout all of the discussion so far, two panel members in particular—Father James T. Burtchaell, professor of theology at Notre Dame, and the conservative attorney James Bopp Jr.—had remained adamantly opposed to fetal-tissue research. They totally rejected the argument that one could separate the act of abortion from the act of using the fetal tissue. And they felt passionately that the mother should not have the right of consent.

Burtchaell had put it very strongly. "When a parent resolves to destroy her unborn, she has abdicated her office and duty as the guardian of her offspring, and therefore forfeits her tutelary powers. Everybody involved in an elective abortion is morally disqualified from deciding how the fetal material should be used, in the same way as a man who has killed his wife is morally disqualified from acting as her executor."

The argument resonated with some members of the panel. After all, it was one thing for a wife to give permission for her deceased husband's heart to be used following an accident like a car crash. But consider a wife who deliberately murdered her husband. In this case few people would argue that she had any right to donate his organs.

But Burtchaell—who saw elective abortion as murder—pressed his argument further. It wasn't just the mother who was disqualified, it was the scientists who wanted to use the material as well. They, too, were accomplices to a crime, and they simply could not morally insulate themselves from the act of abortion by arguing that they had nothing to do with it. Nazi physicians who carried out human experiments in World War II had defended themselves at Nuremberg with the same arguments. "They too believed that they had no say in how those subjects were delivered into their hands."

Burtchaell paused and looked at Judge Adams for effect.

"Let me read to you the words of Professor Julius Hallervorden, who was shipped six hundred preserved brains of 'mercy death' victims for his research in neuropathology. I quote: 'There was wonderful material among those brains: beautiful mental defectives, malformations and early infantile diseases. I accepted these brains of course. Where they came from and how they came to me was really none of my business . . .'

"Hallervorden claimed he didn't agree with what the Nazis were doing, but sought to extract some scientific value out of a tragedy. In his words, 'If you are going to kill these people, at least take out the brains so that the material can be utilized.' We have heard similar arguments in this panel. One researcher said this: 'Abortion is a tragedy, but as long as it occurs, I believe it immoral to let tissue and materials go to waste if it can cure people who are suffering and dying.' "

Judge Adams could see that some panel members were getting very angry at being compared to Nazis. But lawyer James Bopp Jr. developed their case further, arguing that it was impossible to morally insulate yourself from a wrongdoing if you benefited from the results of that wrongdoing on a regular basis. He asked the panel to "Consider a banker who judges narcotics use to be a tragedy, but agrees to accept the proceeds from the local drug network in order to make more capital available for homeowners and small businesses in the area."

The issue, according to Bopp, was one of complicity. Like the banker taking drug money, the scientist using fetal tissue is compromised. "By entering into an institutionalized partnership with the abortion industry as supplier of preference, he or she becomes complicit, though after the fact, with the abortions that have expropriated the tissue for his or her purposes. It is obvious that if the research is sponsored by the National Institutes of Health, the federal government also enters into this same complicity."

Professor Aaron Moscona of the department of molecular and cell genetics at the University of Chicago was furious. He took serious exception to the remarks of Burtchaell and Bopp and passionately resented being compared to Nazi scientists or corrupt bankers. "The Holocaust was not a medical research project to help Parkinson's patients and rescue infants from fatal diseases. It was not scrutinized by peer reviews, examined by NIH panels, publicized by media, open to public questioning, debated in Congress, challenged by the Administration. The Holocaust victims did not board trains out of free will and choice; there were no clergymen, lawyers, ethicists, and social activists urging them to reconsider; they were not advised of their constitutional rights or offered adoption as an alternative.

"At the gates of Auschwitz, no one asked for 'informed consent.' Gas chambers were not a freely elected option to donate skin and hair to make lamp shades and mattresses for the Third Reich. The 'medical experiments' did not involve freely surrendered clumps of embryonic cells lacking neural mechanisms for consciousness and pain."

The atmosphere in the room was explosive. Intelligent men and women who had considered all the rational arguments in good faith were still divided by convictions that they felt passionately.

Moscona continued. "Equating freely surrendered abortus cells with tormented people poisoned with lethal insecticides defies reason and outrages morality. This is not constructive dissent. This might only feed ignorance, inflame passions, and inspire intolerance and extremism."

California Parkinson's Foundation, December 1988

Håkan Widner sat in the library waiting for Bill Langston. He had spent the last few days going over all his notes of the MPTP evaluations. It had been a remarkable experience. After visiting Vacaville prison, he had seen Juanita in Montana, then he had examined George at the California Parkinson's Foundation. Finally, he had been down to Greenfield to see Connie. Now he had to make his recommendation as to which two candidates would go. It would be up to Olle Lindvall to make the final decision, but Widner's choices would be the key factor.

The three remaining candidates were each very sick in different ways. Each desperately wanted a chance for an improved life. Widner looked first at his notes on Juanita Lopez. The reservation in Rocky Boy, Montana, where she had lived for some years with her family, looked nothing like the way he had imagined an American Indian reservation would look. The principal activity there was cattle and horse breeding. Today's Indians were real cowboys. As far as he had been able to tell, Juanita did not participate much in the life of the community. She stayed home watching television, smoking, doing nothing, and getting depressed.

The side effects of her L-dopa medication had continued to plague her. The huge amounts of L-dopa she needed to move had given her too much movement, movement she could not control. Her arms and legs swung around as if under the control of some demon.

Every day was the same. After a night without medication, she woke up feeling rigid and inactive. About fifteen minutes after taking a tablet of Sinemet, she started to move. If she was sitting on a chair, her body started rocking back and forth like a Hassidim praying at the Wailing Wall.

After thirty minutes the rocking became intense and she had to hold on to the sides of the chair to stay seated. Five minutes later the rocking was so violent that she sometimes toppled the chair over.

If she was standing, it was even worse. As the medication kicked in she started a bizarre two-step dance, putting her weight on her right foot and then stepping back onto her left. Then the knees started to rise, the left knee rising when she was on her right foot, the right knee when she was on her left foot. As the dyskinesia worsened the knees would rise almost to her chin. Juanita could do nothing to stop it. She was simply trying to stay on her feet. When she walked she lifted her legs as if stepping over something. When she tried to eat she was often unable to control her hands and she threw the food all over the room.

Eventually, the side effects had become so bad that she was hospitalized for a week in Great Falls under the care of a local neurologist. The purpose was twofold: first to adjust her medication; second, to allow for a full evaluation by Dr. Håkan Widner. As it happened, Widner arrived at the hospital in the evening, the time when Juanita usually had her worst dyskinesias. She knew a little about why he was coming and was very nervous. Nervousness made her dyskinesias even worse. As Widner entered the room she started to writhe back and forth violently. Before he had a chance to introduce himself, he had to run in and prevent her from throwing herself onto the floor.

Juanita's dyskinesias showed how sensitive she was to L-dopa and Widner recommended that her daily dose be lowered. The question he needed to decide was this: When stabilized on a lower dose, would Juanita be a suitable candidate for the fetal-tissue operation? To satisfy Widner, Juanita would have to demonstrate two things: first, that she was reliable and trustworthy; second, that she was physically capable of carrying out a series of clinical tests, such as flipping her hand, walking up and down a corridor, and touching her thumb and index finger together. These clinical tests, carried out over two years or more, would form the objective data that would determine whether the transplants had succeeded or failed.

After carrying out a detailed evaluation, Widner had been satisfied on both points. She was a pleasant and responsible person who was emotionally stable enough to survive the experience, and she was capable of doing the tests. Provided her medication was stabilized, he saw no reason why she should not be a candidate for the experimental surgery.

Widner closed Juanita's file and opened George's. He would always remember the first time he had met George Carillo. George was standing by the door of the examining room, completely frozen like a pillar of salt, with his eyes closed—he discovered later that George had been standing for about two hours waiting for him. As Widner entered the room, he saw George try to raise his eyelids to open his eyes and see the doctor from Sweden who was his only slim hope of a better future. But George couldn't open his eyes.

For eight years, George's life had been a living hell. When he was off medication, his mind was clear and he could think quite normally. The trouble was that his lucid mind was encaged within an unresponsive body. He could perform any movement in his mind, but he couldn't make his muscles do it. Nevertheless, sometimes a loud noise would cause his muscles to react involuntarily—he might blink, for example. Sometimes when George blinked, his eyes slammed shut and he was quite unable to open them again. For long parts of his day George—who could not shout out to someone to help him—was effectively blind.

When, after taking L-dopa medication, George *could* move, he had side effects. The hallucinations were the main problem. Virtually every day for eight years he had had periods, sometimes a few minutes long, in which he saw terrifying visions. Visions of snakes coming at him and attacking his face. Visions of hot flames. Visions of spiders and beetles. Despite having them five or six times a day, each new time George was terrified and shocked by these vivid hallucinations. For him they were completely real, a matter of life and death.

Since he had been released from jail, George had spent a lot of time at Langston's center. He came in frequently for check-ups, mostly with Dr. Jim Tetrud, and was well known to all the staff. George had become a mascot for Langston's research team. He was undoubtedly a marginal figure with a criminal record and a dubious past, but most of the staff at the clinic found him a pretty likable guy. Widner found him likable as well. He had decided to spend as much time as possible with George, to get to know him. Borrowing one of the foundation's cars, donated by Jim Bottomley, they went around the San Jose area, and George showed Widner some of the places he used to hang around.

George's life was grim. Unable to support himself, he existed on welfare. He lived in a series of halfway houses inhabited mostly by unstable and occasionally violent mentally ill patients. George got into fights often—over possessions, food, what to watch on television. Despite his frail appearance and his profound disability, George was strong and had a terrifying temper.

Could George be trusted to participate in a scientific experiment? Widner had few doubts. For five years, George had managed to keep away from drugs, an act that showed great self-discipline. George had told Widner that he hated being dependent on others, that he wanted desperately to be independent again and get a job and an apartment, that he would do whatever it required.

After doing a full evaluation, Widner concluded that George could carry out the necessary tests and was reliable enough to be a member of the study.

Widner closed George's file and opened the last file: that of Connie Sainz. He knew that Bill Langston was hoping he would choose Connie, and he could tell that Connie's family were desperate for her to have the operation. Widner had driven down to Greenfield and had met all her family. Nellie, Connie's mother, seemed to him to be a warm and wise human being. They lived simply in a nice bungalow in Greenfield, with a good-sized and rather beautiful yard where her husband kept canaries.

Nellie and David Sainz had been married in Big Basin, Wyoming, and moved soon thereafter to California. David Sainz worked as a laborer while Nellie raised four children, Connie, Stella, Patsy, and Robert, who were now grown up with children of their own. While Nellie had received very little formal education, all of the children did quite well at school. Stella, for example, became a teacher's aide. Before Connie's tragedy, this family was struggling along. But the tragedy had changed everything. Nellie not only had to care full-time for Connie, she also had to bring up Connie's son, Jason, now 9 years old.

If anyone deserved a break, it was this family. Of all of the MPTP cases, Connie seemed to be the most tragic. While the others had all been long-term drug abusers, Connie had not taken drugs before 1982. She had paid a terrible price for her mistake.

But after a long evaluation, Widner seriously doubted Connie's suitability as a candidate for this scientific study. She was in poor health, and very fragile. Could she survive the trip and the operation? This was not at all clear. If Connie died during surgery, the entire research program could be put in jeapardy.

There were other practical problems. Connie lived permanently in an off state. Since going off L-dopa, she had been frozen. In this state she could not do any of Widner's tests. If he asked her to flip her hand, she couldn't. If he asked her to count to 20, she couldn't. If he asked her to walk unassisted, she couldn't. Off L-dopa, she was not a good candidate, as Widner could not record any objective quantitative data on her clinical condition.

On the other hand, restarting her on L-dopa was dangerous to both her and her family. On L-dopa, this frail woman had been capable of striking feats of strength—running through the house with a kitchen knife in her hand, chasing an imagined intruder. Even tiny amounts of medication might bring back such vivid and intense hallucinations. These hallucinations not only exposed her and her family to risk, but also made her an unreliable candidate for the experiment.

It sounded callous to talk about science when confronted with Connie, but the Lund task force had undertaken this project not simply to treat two MPTP patients but to learn whether it would ever be possible to treat millions of sufferers with Parkinson's disease. They just couldn't take any chances.

Langston came into the library and sat down. "Well, Håkan, what's the verdict?" Widner hesitated. "Well, Bill, the final decision will be up to Olle Lindvall and Anders Björklund, but I think it's pretty clear that there are only two ideal candidates—George and Juanita. They can carry out the standard tests, they are reliable enough, they want the procedure. David, Bill, and Toby aren't good candidates for a variety of reasons. And Connie is just too sick."

Langston hesitated, searching for an argument that might convince Widner to change his mind. It was Connie's case that had led Langston into the collaboration with the Swedes. This operation was her last hope.

"Is there no way that you can include her in your trial for compassionate reasons?"

Widner smiled. He knew what Olle Lindvall would say. Lindvall was a very compassionate man and great clinician, but above all he was a scientist and he wanted these experimental procedures to yield data. Connie could not move and thus could not carry out the required tests. Connie was simply not a good scientific candidate. Additionally, there were other worries.

"Look, Bill. Connie is very sick. If we take her all the way to Sweden for this operation, we might make her worse. We might damage her or she might not survive it. That would be disastrous, and not only for her. Think of the publicity. It might kill the whole program off. But it's probably better for her to wait. She will probably only get one shot at this operation. Isn't it better to try with

George and Juanita and learn how to do the operation as well as possible? If it works for them, then it may be easier to justify doing it for Connie."

Langston knew that Widner was right, but he felt terrible. He would have to tell Nellie that for the time being, there was no hope for Connie. Before he gave in, he made Widner promise that they would do Connie if George's and Juanita's operations were a success.

Third Meeting of the Human Fetal-Tissue Transplantation Panel Consultants to the Advisory Committee to the Director, NIH, December 5, 1988

The first three-day meeting of the panel had heard evidence from dozens of scientists and interest groups, the second two-day meeting was aimed at producing a report. But it had been necessary to call a third meeting. All had struggled with the issues. For Judge Arlin Adams, it had been a problem of weighing one major concern—his objections to abortion—against another major concern—making it possible to do medical research that could improve the lot of thousands of citizens—in a sensible and rational fashion. After hearing all the evidence, he was persuaded that research should be allowed to go ahead under the right set of safeguards.

The element which had tipped the balance for Judge Adams was the risks of the federal government's *not* being involved. If NIH funded such research, it could exercise control and enforce standards through periodic review. Without government funding, Judge Adams was convinced that there would be attempts to use fetal tissue for medical research and for profit which would be totally unsupervised. Pregnant women and fetuses would probably be much better protected if the NIH participated than if it did not.

Other panel members with different attitudes to abortion and scientific research had waged their own internal struggles. Despite the broad nature of the committee, there was an emerging consensus: Research should be allowed to go ahead under a strict set of safeguards. And remarkably, they were virtually the same as those arrived at, quite independently, in Sweden and Britain.

The decision to terminate a pregnancy and the procedures of abortion should be kept independent from the retrieval and use of fetal tissue. Payments (other than for reasonable expenses connected with the retrieval, storage, preparation, and transportation of the tissue) should be prohibited. Pregnant women should be prohibited from designating the recipient of fetal tissue (thus ruling out a daughter's having an abortion to help a parent). Potential recipients of these tissues should be informed of their origin. Procedures must be adopted which accorded human fetal tissue the same respect as other cadaveric human tissue. Regardless of research needs or convenience, health of pregnant women should remain the principal concern in changing any abortion procedures.

The majority of the twenty-one members of Judge Adams's panel agreed and signed his report. Four members did not: James Bopp, Father James T. Burtchaell, Rabbi J. David Bleich of Cardozo Law School in New York, and Georgetown University psychologist Daniel Robinson. Adams presented his report to the NIH advisory panel, who accepted the panel's consensus. On January 9, 1989, James Wyngaarden sent the report to the office of the assistant secretary of health in the new Bush administration.

Then the scientific community waited.

19

The Long Journey

California Parkinson's Foundation, August 13, 1989

Langston watched as Dr. Håkan Widner did a final evaluation of George. In a few hours George would be in the air, bound for Sweden. If everything went well with him, Juanita would fly to Sweden in about a month.

A great deal was at stake. For over a year and a half, a federal funding ban on fetal-tissue transplants had effectively killed off U.S. research in this area. Despite the outcry among doctors and patients, despite the recommendations of Judge Adams's fetal-tissue panel, the Bush administration had held firm. Passionately anti-abortionist, they saw themselves fighting a holy war. It was a war they seemed to be winning. They had cut off federal funding for abortion counseling. They had openly declared their intention to appoint pro-life justices to the Supreme Court as soon as some of the very elderly incumbents, like Justices William J. Brennan Jr., Thurgood Marshall, and Byron

White, retired. The *Roe* v. *Wade* decision—the legal basis for a woman's right to choose—would then be in danger.

While *Roe* v. *Wade* survived, it was still possible to do fetal-tissue transplants in the United States using private money. Eugene Redmond's group at Yale and Curt Freed's in Colorado had somehow managed to put together privately funded programs. If *Roe* v. *Wade* fell, their fate would be left in the hands of state lawmakers. If Connecticut's or Colorado's politicians chose to make abortion illegal, then those programs would have to shut down or move to a state which continued to permit abortion.

Both groups had inquired about doing fetal-tissue transplants on Langston's MPTP patients. The idea was very tempting. It was much simpler to fly a patient to Colorado than halfway around the globe to Sweden. But despite the logistical difficulties, Langston had no doubts that George and Juanita would be getting the best possible chance at Lund. No group anywhere in the world could compare in experience to the Swedes—neither the Yale nor the Colorado group had been grafting human patients when Langston met Björklund on the bus in Belgium.

The Swedes had now done a total of four human cases. The two female patients who received fetal-tissue grafts in 1987 had continued to improve modestly. In April and May of 1989, the Swedish team had carried out two more fetal-tissue transplants. The first of these was a 49-year-old man who had been diagnosed with Parkinson's disease in 1977. His main problems were rigidity and bradykinesia in his right arm. This patient received a unilateral fetal transplant, using tissue from four fetuses, into the left putamen in an attempt to compensate for the symptoms on his right side. The second patient was a 59-year-old man who had been diagnosed with Parkinson's disease in 1982. His main problem prior to surgery was left-sided rigidity and bradykinesia. He received tissue from four fetuses grafted into the right putamen, to compensate for symptoms on his left side. Prior to surgery, both patients had suffered with fluctuations between on and off states for three or more years.

It was much too early to know whether these transplants were successful. But it was reassuring to Langston that the team had successfully carried out the surgical procedure, with the complicated planning and logistics of procuring the fetal tissue, on two more patients.

While the Swedish group had only carried out four human cases, this clinical work was founded on decades of basic research in animals. In this respect they were unique. If he had needed a fetal-tissue transplant for himself, Langston would have it done in Sweden.

Langston had been amazed at Håkan Widner's dedication and persistence. For a year and a half, Widner had regularly come out to California and spent time in prisons and halfway houses with criminals and drug addicts, all in the name of science. He had done an incredible job setting up the scientific protocol and getting the vital baseline data on what George and Juanita could do before surgery. If George and Juanita got better, Langston thought, he really had the Swedes in general and Widner in particular to thank.

The meeting on the bus in Belgium had given birth to a great idea, but putting the idea into practice had taken years of planning. First there was the problem of paying for the operations. Langston had gone to all four of the Parkinson's disease foundations, asking each for a contribution of ten thousand dollars. Despite their professed interest in the project, only one foundation had contributed funds—the New York–based Parkinson's Disease Foundation. The others all found reasons for not supporting the research. Having been burned by the adrenal tissue fiasco, the United Parkinson's Foundation argued that the transplants were premature. The American Parkinson's Disease Association turned Langston down, saying they were worried about liability issues if something went wrong. The Miami-based National Parkinson's Foundation group refused Langston's request because, they said, their board was concerned about possible negative publicity.

The Dystonia Research Foundation of Canada agreed to fund the Canadian part of the project—the PET scans in Vancouver.

The bulk of the money for George's and Juanita's operations came from Sweden. The Tricentenary Fund of the National Bank of Sweden, the Swedish Medical Research Council, and the Westerström Foundation all made significant contributions. The biggest component—the hospital costs—were absorbed by Lund hospital, which agreed to treat George and Juanita as if they were Swedish nationals and therefore entitled to free medical care.

Then there had been the ethical review. The protocol had to be passed by human subjects review boards, not only in Lund but also at Santa Clara Valley Medical Center.

Everything had been done carefully by the rules. But Langston was desperately worried about the fallout should something go wrong. These patients were by now famous. Among neuroscientists they were celebrities, known by their first names. Scientifically speaking, George and Juanita were precious resources, not to be squandered lightly. Their personal tragedy had led to a continuing scientific revolution; if something happened to them it would be both a personal and a scientific tragedy.

If anything went wrong, if George or Juanita suffered any ill effects from the operations, if either of them died, the Swedish-California team would be severely criticized. People would say that the operations had been premature, that they should not have risked the lives of such valuable patients, that they had acted unethically.

If the popular press took hold of the story, the situation might be even worse. They were about to send a pair of former drug addicts with criminal records halfway round the world to Sweden, to receive tissue transplants from the brains of aborted fetuses. Tens of thousands of dollars would be spent in this effort to restore George and Juanita, and for what? Langston could just imagine the headline: SWEDISH SCIENTISTS USE ABORTED FETUSES TO SAVE DRUG ADDICTS.

While George rested, Langston and Widner reviewed the elaborate travel arrangements. The biggest issue was George's medication. If George was allowed to take L-dopa, there was a very good chance that he would have terrible hallucinations and wild dyskinesias during the trip. If he started seeing snakes and fire at thirty-five thousand feet over the Atlantic, it would be a disaster. Langston had nightmares of the pilot having to make an emergency landing at London's Heathrow Airport. On the other hand, if George was taken off all medication, he would be unable to move at all. He would be unable to go to the bathroom and to eat and drink during a fifteen-hour journey.

Håkan Widner had traveled with George several times to Vancouver for PET scans and had learned how to control George's symptoms by giving him tiny amounts of L-dopa when he needed it. He planned to keep George off medication for the short trip to Los Angeles and then give him several small doses during the eleven-hour trip to Europe. This would not eliminate the risk of hallucinations, but it at least reduced them.

Langston said goodbye and wished them both luck. If all went well, he would join them in Lund in about a week's time. As he watched them walk out into the parking lot of the California Parkinson's Foundation, Langston thought that Widner and Carillo made a strange pair of traveling companions—the stocky, blond, blue-eyed Swedish doctor and the short, skinny, dark Hispanic patient. George even walked with a Chaplinesque shuffle.

The journey was a long and complicated one. Lund is in the southern part of Sweden, nearer to Copenhagen than to Stockholm. So first they would fly to Los Angeles. Then from LAX they would make a long polar flight to Copenhagen International Airport in Denmark. There a Hovercraft would take them the short journey to Malmö in Sweden, which was only a short car ride from Lund.

Copenhagen, Denmark, August 14, 1989

Widner watched George eat his breakfast. The L-dopa was wearing off and George had some difficulty lifting the spoon to his lips, but he had passed the night without a major hallucination. In fact, everything had gone remarkably well. George had consumed a double order of steak for dinner and watched the movie. He seemed to be taking full advantage of SAS's in-flight service.

In thirty minutes they would be landing at Copenhagen airport, where there would be a vehicle to take them to a Hovercraft. The Hovercraft trip would take about forty-five minutes. After clearing customs, there would be the friendly faces of Olle Lindvall and Patrik Brundin waiting to greet them and take them to the hospital.

Lund, Sweden, August 21, 1989

During the first week in the hospital, George was given a complete medical workup and a full battery of tests. When he wasn't being scanned or having blood drawn, he had been entertained by the two youngest members of the team, Drs. Håkan Widner and Patrik Brundin. They had rented him dozens of violent videos (at his request)—*Scarface* was his favorite—and brought him hamburgers to make him feel at home. They had even taken him out in a wheelchair to see the Malmö festival—a Swedish cultural event. George particularly liked a street band playing jazz and rock on homemade instruments. He had bought a tape of the band, and played it often on a borrowed tape recorder.

Still, George felt lonely and a little scared. The first operation was scheduled to take place tomorrow, and he wasn't quite sure what was involved. He had heard that Dr. Langston had arrived in Lund and would be over to see him shortly.

Langston was in the small library of the Lund neurology department meeting with the transplant team. Olle Lindvall was on his feet talking. On the board, he had drawn a cutaway picture of the brain and was explaining how the neurosurgeon would perform the transplants into George's brain.

"The neurosurgeon, Stig Rehncrona, is going to try to do the entire operation through a single small burr hole. Imagine the hole as a fulcrum. By rotating the injection cannula between insertions, he can reach a target in the caudate nucleus and the putamen. We hope this will work. If it doesn't he will have to make two burr holes."

"How long will the procedure take?" asked Langston.

Lindvall laid out the schedule. George would be put under general anesthesia at 8:00 A.M. Next, the stereotactic frame would be bolted to George's skull. Then George would be taken to radiology, where they would carry out a CAT scan and do the calculations that Stig Rehncrona needed for his operation. The implantations would begin around 1:00 P.M. As they were hoping to implant at least four sites, this should take about three hours.

Several changes had been made to the procedure since the first two fetal transplants were done in 1987. The cannula they now used was much smaller—1 millimeter instead of 2.5 millimeters in diameter. A series of fetal grafts in rats had produced much better cell survival using the smaller cannula. They hoped to increase cell survival by reducing to a minimum the delay between the procurement of the tissue and its implantation. They would be using only fetal tissue harvested from abortions on that day.

Another big difference was that they planned to implant fetal tissue into both sides of George's brain, making George the first patient in history to receive a bilateral brain graft. Today's operation would implant fetal tissue into the right side of George's brain. Another operation, in a week to ten days, would repeat the procedure for the left side. This would violate a golden rule of neurosurgery, which dictated that it was dangerous to operate on both sides of the brain within a short time. It would also greatly increase the cost and logistical complexity of the procedure. But the team had decided it was essential to give George a chance of a normal life. George's parkinsonism was so bad on both sides that if a unilateral graft worked, George might be turned into a freak, able to move on one side and frozen on the other.

"What's the likelihood of getting enough fetal tissue?" asked Langston.

"That's the big unknown," Patrik Brundin chipped in. "We can't guarantee there will be enough suitable tissue. We hope it will be okay, but there is no guarantee that we will get enough. Our calculations show we need material from at least three fetuses; otherwise, the yield will be too low to make any difference."

"What happens if you don't get enough?" Langston asked anxiously.

"Then we don't do the operation. George will be sent back to his bed and next week we try again."

Surgical Unit, Lund Hospital, August 22, 8:00 A.M.

George Carillo slept through the most important day of his life. While he slept under anesthesia, the team began assembling the stereotactic frame. The operations on George's and Juanita's brains would not be open procedures like those carried out in Mexico City, but *closed* procedures, performed blind through a hole in the skull. Being less invasive, with less trauma and less risk of infection, closed operations are safer in principle. But closed procedures depend totally on stereotaxis. Without the stereotactic frame the surgeon has no way of orienting himself in three-dimensional space.

After bolting the stereotactic frame to George's skull, nurses wheeled him to radiology for a CAT scan. After his head was positioned precisely, the million-dollar machinery began to work, taking hundreds of X-ray pictures of his head which the computer integrated together to provide a three-dimensional picture of George's brain. Behind a large window, neurosurgeon Stig Rehncrona looked at a high-resolution screen on which the radiologist could conjure up a picture of any slice of George's brain viewed from any angle. Rehncrona would not be able to use the CAT scan to navigate during the operation; he would be dependent on the stereotactic frame. All of the measurements and calculations that he was about to do had to be made with reference to the stereotactic frame.

The radiologist concentrated on the image of the steel bolts of the stereotactic frame. Carefully, she moved a cursor over the image of each of the bolts in turn and clicked on her mouse, entering the coordinates into the computer. She did this in the *x* axis, the *y* axis, and the *z* axis. These coordinates were the reference points for all the calculations that would follow. Any measurements that would be made, such as the angle of attack or the depth of implantation, would be made relative to these points.

Now Stig Rehncrona began to talk, asking for one view, then another. He wanted to plan a route into George's striatum that would be direct but avoid major blood vessels. His targets were the caudate nucleus and putamen. The most direct path to those targets was from from the back of George's skull. After experimenting with different angles, Rehncrona decided on a path. He carefully wrote down the parameters of the operation: the entry point into George's brain, the angles at which the cannula would be inserted, and the depth of that insertion, so that he would arrive at his targets in the putamen and caudate nucleus.

Tissue Culture Lab, Obstetrics and Gynecology Building, 9:10 A.M.

Patrik Brundin had just received the first aborted fetus of the day. After carefully sterilizing all of his instruments, he placed the tissue under the microscope. Like most abortions in Sweden, this one had been done using the suction method, which, while quick and relatively painless for women, fragments the fetus. Under the microscope, Brundin saw a jumble of white, translucent pieces of the fetal cadaver. His task today would be to try and reassemble the fetus, like a jigsaw puzzle, so that he could identify the area that contained the substantia nigra. Then he would take that part and remove all the tissue which was fated to become bone, skin, and cartilage. It required enormous skill—the entire fetus is only the size of a fingernail and the substantia nigra the size of a pinhead. But Brundin had done this thousands of times with rat embryos, which look very similar.

The first fetus looked good. The suction hadn't damaged the fetus too much and Brundin knew he would have little difficulty finding the correct part and dissecting out the substantia nigra cells; it would take him about fifteen minutes. He stored the fetus in a carefully balanced solution and waited. He would need at least three and preferably four dissectable fetuses to make the operation worthwhile.

While Brundin waited for more fetal tissue, Olle Lindvall was pacing out one more time the route from Brundin's lab on the third floor of the obstetrics and gynecology building to the operating room on the seventh floor of the main hospital building. It involved a long walk and an elevator ride. Like most teaching hospitals, Lund was a very busy place, with patients and doctors moving about at all times. Lindvall had nightmares that the person carrying the tissue might get lost or be bumped in the elevator, causing him to drop the precious cells. Lindvall had decided that when Patrik Brundin carried the tray, he would be surrounded by foot soldiers—Anders Björklund, Bill Langston, and himself—to prevent his being jogged. Moreover, Brundin would wear a sterile surgical gown, mask, and gloves to impress bystanders to keep out of the way.

By 10:00 A.M. the radiology was finished and George was wheeled into the OR, where the anesthesiologists continued to monitor his vital signs. The nurses checked that everything was ready for the operation, and Stig Rehncrona began adjusting the stereotactic frame.

The second fetus that Brundin received had been good, and he had been able to locate the substantia nigra. But the third and fourth had been far too damaged. The fifth had been promising, but much too small. There was one more fetus expected soon. If this was inadequate, then the operation would have to be postponed. Material from three fetuses was the bare minimum needed to get any effect.

By 11:45, when the sixth fetus arrived, Brundin was beginning to get worried. He began his examination. Proceeding cautiously, he reassembled the pieces into the recognizable form of a fetus. Yes, this looked promising. As he worked, his confidence grew. Yes, it looked like they might be all right.

Brundin took a new set of sterilized instruments and began the dissection of the first fetus. He located the substantia nigra region and began dissecting out all the tissue which was fated to become bone, skin, and cartilage. Each fetus took between fifteen and thirty minutes.

After finishing each dissection, Brundin immersed the tissue fragments in a 0.1 percent solution of trypsin to break down the connective tissue holding the structures together so the cells could later be dissociated into a suspension. The optimal time for this immersion was twenty minutes. Any more and the trypsin might damage the fetal cells.

By 2:30 Brundin had finished dissecting his third fetus. He removed the last mass of dissected fetal tissue from the trypsin, washed it, and placed it in one of several glass phials containing saline solution. He then inserted the phials in a metal capsule, which in turn was placed in a metal box that he wrapped in a surgical cloth. Brundin picked up the box gingerly and, surrounded by his escorts Olle Lindvall, Anders Björklund, and Bill Langston, set off for the operating room. Fully dressed in surgical garb, Brundin looked like a high priest involved in some bizarre ritual. And indeed, there was something reverential about what they were doing. What he carried in his hands was immensely precious. If today's operation worked, the cells would become part of George's brain and still do what they were designed to do—make dopamine.

The procession made the journey without incident and handed the tissue to the OR nurse, who in turn handed it to Håkan Widner, who anxiously waited in the OR. Widner's job today was to time critical steps in the transplant and to assist with the preparation of the tissue. Patrik Brundin went into the changing room and changed quickly into a new set of surgical clothes.

Back in the OR, Stig Rehncrona picked up the electric drill and began making a tiny burr hole in the top of George's skull. This was the entry point to George's brain—with luck, the only hole Rehncrona would have to make in his skull that day.

Through an observation window, Langston, Lindvall, and Björklund sat anxiously watching the unfolding drama. Stig Rehncrona checked the settings on the stereotactic frame. Håkan Widner picked up a stopwatch and checked that it was working correctly. The OR nurses checked their instrument tables.

Patrik Brundin was now in the operating room, sitting at a table a few feet to Widner's right. Each fetus he had dissected contained roughly a million dopaminergic cells. He now had to prepare them for Stig Rehncrona. Large, solid pieces of tissue are impractical for a fine stereotactic cannula, so they had to be dissociated into smaller pieces—but not too small. Their animal research had shown that a single-cell suspension produced a very poor overall survival rate.

Brundin began carefully drawing up the mixture of clumps of fetal tissue and saline solution into a syringe. Deliberately, he began pushing and pulling the plunger of the syringe, using the mechanical energy to dissociate the cells into clumps of one hundred cells or so, the optimum size for today's procedure.

At 2:49 P.M., having set the stereotactic frame to the correct angles, Rehncrona carefully introduced the cannula—a rigid hollow wire—through the small hole in George's skull. Knowing not only the bearing but the depth of his targets, he planned to make three implantations in George's putamen and one in George's caudate nucleus. The PET scan had revealed that fluorodopa uptake in these areas was markedly reduced.

Like most neurosurgeons, Stig Rehncrona had incredibly steady hands. He held the guide cannula perfectly still as he began gently guiding it to its target. The cannula slipped gracefully between the folds of delicate nerve tissue. There are enormous risks to this procedure if it is not done correctly. One false move and George's brain could be damaged for life.

The 1.5-millimeter guide cannula, which Rehncrona had designed himself, was actually a hollow tube with a wire stylet placed inside it filling the tube so that it did not get clogged with brain tissue as it was pushed in. Once the tip of the cannula had reached the putamen, however, Rehncrona withdrew the stylet through the middle of the instrument. Then he introduced a second, finer, cannula, 1 millimeter in diameter, through the guide cannula and pushed on into the putamen itself. Now a direct channel existed through which fetal cells could be transferred to George's brain.

Taking the syringe of fetal cells that Patrik Brundin had prepared, Stig Rehncrona carefully and slowly injected it into the second cannula. To ensure that the tissue was well distributed, they had agreed to deposit the fetal suspension in eight portions, 2.5 microliters of fetal-cell suspension at a time, which they would attempt to spread out in the putamen and caudate.

As Stig Rehncrona squeezed the first 2.5 microliters through the cannula, Håkan Widner started his stopwatch. It was Widner's job today to pace Rehncrona. The speed with which the cells were deposited into the brain was critical, for a number of reasons. First, they did not want to damage the cells by injecting them too rapidly, so each deposit was spread out over twenty seconds. Second, they had to allow the cells to settle into the right site, and not be sucked out after being deposited when the instrument was withdrawn. Through their careful animal work, the Lund team had learned that it was necessary to leave the instrument in place for about ninety seconds before starting to withdraw it. During this time, the fluid in the suspension was absorbed into the surrounding brain tissue and the fetal cells settled into their target location.

After a ninety-second wait, Stig Rehncrona withdrew the cannula just 1.5 millimeters and, still in the putamen, repeated the process, injecting the second 2.5 microliters of fetal-cell suspension. In this manner, working slowly and deliberately, the fetal material for this first implant into the putamen was deposited in eight closely related sites in the putamen. By 3:15, the first implant was finished.

Now Stig Rehncrona adjusted the stereotactic instrument to a new set of coordinates and repeated the procedure using the same cannula, making a second implant in the putamen. Each implant deposited eight packets of 2.5 microliters of cell suspension to a target region. At 5:15, he finished the final implant. Everything had gone as planned. Rehncrona had made three implants into the putamen and one into the caudate. A total of 80 microliters of fetal-tissue suspension had been transplanted into George's brain.

Now that the valuable cells had been transplanted, the team had to ensure that George's body would not kill them off. George had been started on cyclosporine, an immunosuppressant drug, a few days before surgery, to guard against possible rejection. Now, to ward off infection, George was given antibiotics.

Anders Björklund had set up a microscope just outside the OR to do a final check. He wanted to do a last-minute examination of the fetal tissue to estimate what proportion of the cells were still alive. The processes of abortion, dissection, transport, and dissociation had all killed off cells. Björklund mixed 1 microliter of the cell solution with two fluorescent dyes, acridine orange and ethidium bromide. Dead cells cannot shut out the red ethidium bromide that sticks to DNA in the cell nucleus, whereas living cells accumulate the yellow acridine orange dye. What this means is that under the microscope, dead or dying cells light up as red spots and living, healthy ones shine a fluorescent green. Björklund counted the relative numbers of some five hundred cells and found that roughly 70 percent of them were green and therefore alive. This was good news. It meant that possibly 70 percent of the implanted cells were still alive as well.

Björklund realized that many more cells would die inside George's brain over the next few days, especially if they didn't get the appropriate nourishment from the surrounding environment. If their careful animal research was any guide, in a week George would be left with perhaps 10 to 20 percent of the cells Patrik Brundin had started with. But with material from three fetuses, that ought to still be enough to produce a clinical effect.

By 7:00 P.M., George was wide awake, with Stig Rehncrona, Ole Lindvall, Håkan Widner, and Bill Langston by his bedside. A full postoperative checkup revealed that George had no major problems, apart from vivid hallucinations—which many Parkinson's patients experience following general anesthesia. George did not appear to have any neurological complications from surgery.

The day after surgery Langston spent some time with his most famous patient before returning to California. Langston and Lindvall examined George and chatted with him for a while. While George seemed as grumpy as ever, Langston felt cautiously optimistic. George seemed to have come through the first test unscathed.

Fax from Dr. Håkan Widner to Dr. Bill Langston, August 28, 1989

Things have settled down a little here. As I think I told you on the telephone, we had a lot of trouble with the weekend and night staff caring for George. To put it bluntly, they are terrified of him and at one point last Sunday, they refused to care for him unless he was tranquilized.

The problem is that they have never encountered anyone like George before and treat him as if he were a demented bum. I have explained until I am blue in the face about his parkinsonism and about how he lives in a slow motion world, but they don't seem to have understood. They don't understand how hard George is trying all the time to do things. So they treat him like an idiot and prod him and cajole him into doing things. As you know, George hates to be treated like this and gets very angry. Things came to a head last Sunday when George was lying quite still hallucinating and a night nurse leaned over him and touched him to get him to do something. George snapped back into consciousness and grabbed her. You know how strong George is. Well, anyway the night staff were absolutely terrified. When I got there they refused to care for him.

At one point, they wanted to give George neuroleptics to calm him down. I of course told them that this would kill someone with advanced parkinsonism like George. They asked about other drugs. So I said to them, "If you were pregnant would you take drugs?" " 'Of course not,' they replied. "Well, George has just had a fetal-tissue transplant. The cells in his brain are fetal cells a few weeks old, as sensitive to drugs as a first trimester 'baby.' If we give George tranquilizers we risk hurting those cells." This argument got through and we began to discuss constructive ways of treating George with proper respect so that he would not explode with anger.

On Monday, the day staff came in and everything was fine again. Lene has no trouble handling George and fully understands his needs and problems, but unfortunately, Lene can't be there all the time.

Best Wishes,
Håkan

Fax from Dr. Håkan Widner to Dr. Bill Langston, September 16, 1989

As I think you heard, George's second operation on August 29 was something of a disaster. Poor George was under general anesthetic with the stereotactic frame in place, but we were unable to procure sufficient fetal tissue. We had no option but to wake him up and send him back to his room. He was, as you can imagine, a little disappointed. Anyway, things went much better last week on September 5. We managed to get four viable fetuses and made four injections to the left side of George's brain, transplanting about 80 microliters of tissue suspension.

Patrik and I have been trying to keep his spirits up. He knows that it will be some time before there are any results, but I think he is a little disappointed that nothing seems to have changed. I will be flying back to California with George on September 19 and will see you then before I bring Juanita back to Lund with me. Her operations are scheduled for October 16 and November 2.

Best Wishes,
Håkan

Fax from Dr. Håkan Widner to Dr. Bill Langston, November 10, 1989

Juanita's stay here in Lund has been uneventful and she has been a delightful patient. She has grown quite close to Lene and the pair have played cards every day and talked about their families and children. Juanita has also grown very fond of our soft cheese here in Sweden and wants to take some back with her.

The only big problem we have had with Juanita is her smoking. She smokes a lot and smoking is not allowed in the hospital rooms. Unfortunately, because she is on cyclosporine she is not allowed to leave her room to go outside for a smoke. So in the end we had to bend hospital policy and allow her to smoke in the room.

The smoking was almost certainly behind the attack of pneumonia she got following her first transplant operation. As you know it's very common for parkinsonian patients who are heavy smokers to contract pneumonia following general anesthesia. This is something we need to watch with her.

A bigger problem that I see is the difficulty of finding a vein. Juanita has no good veins for inserting a needle. In her first operation, it took the anesthesiologist over an hour to find a vein capable of holding a needle. This may be a big problem in taking blood samples to monitor her cyclosporine. It may be something the doctors and nurses at Rocky Boy have difficulty with.

Anyway, I'll be bringing her home tomorrow and will report further once I am in the U.S.

Best Wishes,
Håkan

Discharge Summary from Department of Neurology, Lund University Hospital, Lund, Sweden, November 11, 1989

Patient: Juanita Lopez
Diagnosis: Secondary parkinsonism 332 D, due to MPTP-toxicity 968A.
Treatment: Stereotactic implantation of mesencephalic neurons to the left and right putamen and the left caudate nucleus 0234 which took place on October 16 and October 31.
Course: The surgical procedures were totally uncomplicated and the postoperative course totally uneventful.
Prescription: The patient is transferred back to the U.S. for further evaluation by Drs. J. William Langston and James Tetrud, California Parkinson's Foundation, and by Dr. Håkan Widner on an intermittent basis. The result of

the operation is very gradual and slow, and the patient will be under evaluation for about a year. There will be several "one-dose" tests approximately every 2nd month and if possible timed test measurements. A PET scan will be made in Vancouver, Canada, after 4–6 months. Electrophysiological tests will be made in California after 4–6 weeks.
Medication: IT IS IMPERATIVE THAT HER ANTIPARKINSONIAN MEDICATION REMAINS UNCHANGED. ANY TAMPERING WITH SINEMET WILL INTERFERE WITH THE INTERPRETATION OF THE OPERATION. CONTACT DR. WIDNER OR DR. TETRUD BEFORE ANY CHANGES ARE MADE.

THE IMPLANTED FETAL TISSUE IS UNDER ACTIVE DEVELOPMENT FOR AT LEAST A YEAR FROM SURGERY AND SHOULD BE REGARDED AS SENSITIVE TO DRUGS AS A PREGNANT PERSON. USE THE SAME CAUTION WHEN PRESCRIBING DRUGS AS TO PREGNANT WOMEN.

NOTE! ALL DRUGS AFFECTING DOPAMINE TRANSMISSION (NEUROLEPTIC, DA ANTAGONISTIC DRUGS ETC.) ARE NOT TO BE USED.

20

Out on a Limb

Office of Director, National Institutes of Health, Bethesda, Maryland, November 1989

The newly appointed director of the NIH, Bernadine Healey, held the letter from Louis Sullivan, secretary of health and human services, in her hands. After sitting on the NIH report for nine months, Sullivan was at last officially making a ruling on Ed Oldfield's request to carry out fetal-tissue transplants in humans.

The temporary moratorium had been in place for twenty months. A blue-ribbon panel had considered the issues and advised the NIH advisory committee, on which Healey had been a member. They in turn had advised the new Bush administration. It had been a very thorough inquiry and it had come up with conclusions almost identical to similar bodies in European countries. They had recommended that research be allowed to go ahead under strict ethical guidelines. Now, after months of dithering, Sullivan had announced the administration's final decision on the matter: "After carefully reviewing all the materials, I am persuaded that one must accept the likelihood that permitting the human fetal research at issue will increase the incidence of abortion across the country." The temporary funding ban would now be permanent.

It didn't matter that the panel had specifically considered this issue and concluded the opposite: that fetal-tissue transplants would not affect the incidence of abortion. In fact, nothing the panel had said seemed to have made any difference at all. This was an ideological decision, pure and simple. As a servant of the administration Bernadine Healey would now have to defend a policy she didn't believe in. It wasn't a total ban on all fetal research supported by federal funds. The ban only applied to the transplantation of human fetal tissue from induced abortions. Thus basic research could be done on material from induced abortions. And material from spontaneous abortions and miscarriages could be used for transplantation. As most scientists had argued that it would be unethical to transplant damaged, perhaps infected, tissue from spontaneous abortions and ectopic pregnancies into patients' brains, this was of little help. So while most basic laboratory research on fetal tissue could continue, for patients with Parkinson's disease the news was bad.

And the situation might get much much more difficult. A number of states were challenging the *Roe* v. *Wade* ruling of the Supreme Court. The makeup of the court itself had become critically important. A number of the justices who supported a woman's right to choose an abortion were very elderly and would soon be stepping down from the Supreme Court. If Bush was reelected for a second term, they would almost certainly be replaced by conservative Bush appointees. It was entirely possible then that *Roe* v. *Wade* would be overturned.

If *Roe* v. *Wade* was overturned, several states, perhaps a majority of states, would be within their rights to ban abortion. This would make human fetal-tissue transplants from induced abortions illegal. If *Roe* v. *Wade* was overturned in the Supreme Court and the state of Connecticut, for example, banned abortion, then Eugene Redmond would have to stop his work or move to another state.

California Parkinson's Foundation, Spring 1991

Bill Langston was feeling gloomy. He had just finished examining George Carillo and could find no evidence of improvement. If anything, George's parkinsonism was worse than before surgery, as measured on the standardized scales they had been using: the time taken to do hand flips, the number of finger taps in a given interval, the time taken to walk up and down a corridor. If Langston was gloomy, George was even more depressed. George's hopes had been high. He had been halfway round the world and submitted to three major surgeries, and for what?

On the political front, things looked bad as well. Langston had appeared on a KQED panel discussion about fetal-tissue transplants with a representative of the right-to-life movement, who had tried to compare him to Dr. Mengele, the Nazi physician who experimented on Jewish prisoners in the Second World War.

The only good news was Juanita. Within three months she had noticed some improvement in her voluntary movement, although her sensitivity to L-dopa increased. After much debate, Widner and Langston made a critical decision to reduce her L-dopa medication, even though the research protocol forbade it. Thankfully, she had continued to improve. Håkan Widner had visited her in Montana and carried out a battery of objective tests to rate the severity of her parkinsonism, and she showed a slight improvement. There was nothing conclusive or indeed that impressive so far, but at least there was measurable progress.

It was critically important that the MPTP cases achieve success. While Langston had been careful to keep the details of George's and Juanita's operations out of the media, Parkinson's groups knew about them and a lot of patients were putting their hopes on this line of research. The various patient groups and the pro-choice lobby had regrouped and were fighting back against the Bush administration. For them, a successful outcome with George and Juanita would provide powerful ammunition to fight the Bush ban. Representative Henry Waxman of California planned to introduced a bill to reverse the federal funding ban on fetal-tissue transplants. The problem was that unless they got at least two-thirds of the senators and representatives to vote for this bill, George Bush would be able to veto it.

21

The Turning Point

California Parkinson's Foundation,
December 1990

Håkan Widner was back in San Jose to see how George was progressing, now more than a year after surgery. Before returning to Sweden, he would go up to Montana to evaluate Juanita.

Widner decided to do a complete review of all of George's clinical tapes. Every two weeks for a year following surgery, Dr. Jim Tetrud had asked George to come in and run through a battery of tests. The first tape, taken in October 1989 immediately after his arrival home from Sweden, showed George with a shaven head looking decidedly the worse for wear. The twenty-six tapes, each one of which was about an hour in length, made very boring viewing, even for a neurologist. In each tape, the same thing happened over and over again: George was shown doing hand flips, foot tapping, walking up and down the corridor, etc. Langston had seen most of the tapes individually, but had never been able to face what Widner was now doing.

After five hours, Widner noticed something. By watching

the tapes in sequence, he had picked up a very subtle and slow improvement in George. The improvement was so slow, so insidious that it was difficult to detect when looking at any two or three tapes in sequence. But it was definitely there.

He set up the series of tapes in chronological order again, but this time he cued each tape to the foot-tapping exercise. In quick succession he fed each one of the tapes into the player and let it run for thirty seconds through the foot-tapping exercise. Shown this way, simulating time-lapse photography, the tapes showed George tapping his feet at two-week intervals over the past year. It was very subtle, but there was no mistaking the change. In the later tapes, George could lift his foot higher, he could move more easily. It was a clear-cut improvement. George was getting better.

Widner called Jim Tetrud and Bill Langston into the library where he was viewing the tapes. They watched and agreed that there was a progressive improvement in George's ability to move. The group began to get very excited. Perhaps, Langston speculated, the delay could be explained by the severity of George's condition. Because George had lost so many dopamine cells, probably more than all but the most severe Parkinson's patients, a transplant might not have an immediate effect. Before any clinical effects could be seen, the cells would have to develop and form connections with surrounding brain tissue, and make more and more dopamine. In other words, there was a threshold effect not seen in the Swedish idiopathic Parkinson's disease patients. Jim Tetrud reminded them all that the implanted cells were a few weeks old at the time of surgery and that very little was known about such fetal tissue. In all probability the cells were not making much dopamine in the fetus and needed to develop and grow before producing clinically significant quantities of dopamine. And as George's brain before surgery was so depleted, he needed lots of dopamine.

Over the next few months George continued to improve

slowly but steadily. He began to walk with a natural arm swing. Some expression returned to his face. He still had problems with L-dopa-induced hallucinations, but things were looking up.

Back in Lund, Widner could report that not only George but also Juanita was making steady progress. It seemed as if the grafts had established themselves and were making a functional difference to both patients' lives.

The news of the third and fourth Parkinson's disease cases that had been grafted in Lund in the spring of 1989 was also good. During the first year following the operation, both patients' symptoms had ameliorated significantly; the symptoms of rigidity and bradykinesia were less severe and the time spent in the off phase was reduced. For both patients, a single dose of L-dopa lasted longer than before surgery. Moreover, PET scans had revealed that the grafts were alive and functioning well in both patients. The scans showed an increased uptake of fluorodopa in the grafted putamen in both patients.

During the second year following surgery, one of these patients continued to improve dramatically, and after thirty-two months, it had been decided to withdraw L-dopa therapy altogether. PET scans showed that in this patient fluorodopa uptake continued to increase in the grafted putamen, as contrasted with his ungrafted putamen, which showed no change. The other patient only made modest improvement in the second year. His PET scans revealed that fluorodopa uptake was unchanged in his grafted putamen but actually decreasing in the putamen on the ungrafted side.

Taken together, these results were very encouraging. They showed that fetal tissue survived, grew, and exerted functional effects, even when an ongoing disease process continued to operate in the ungrafted half of the brain.

House Subcommittee on Health, April 1991

Representative Henry Waxman appeared pleased with the way things were going. He had been holding hearings in furtherance of his bill, part of which sought to overturn the Bush funding ban on fetal-tissue research. The witnesses had been carefully chosen to provide the maximum impact. Scientists had testified to the encouraging results that were being obtained in fetal-tissue research, such as the third and fourth Swedish patients with Parkinson's disease. But his star witnesses had been the Reverend and Mrs. Guy Walden.

Guy Walden was a Baptist pastor from Houston, Texas, who was fervently against abortion. What made him exceptional was that he was in favor of allowing fetal-tissue transplants. In fact, he had allowed researchers to implant fetal tissue into one of their children.

Three of Mr. and Mrs. Walden's six children had been born with a rare genetic disease called mucopolysaccharidosis type I. In this invariably fatal disease the baby is born with an enzyme deficiency—it had claimed the lives of two of their children, Jason, at age 8, and Angie, at age 7. When they learned that another of their children, baby Nathan, was threatened, Terri and Guy Walden decided to take action. They learned of an experimental new treatment in which bone marrow from an aborted embryo is implanted into the baby while it is still in the womb. Before making the decision, the couple agonized and searched the Scriptures to find guidance and finally found what they took to be a reference to the subject in Genesis. "When God made Eve out of Adam's rib," said Guy Walden, "It was clear to us that God was not opposed to tissue transplants."

The procedure seemed to have a positive effect, and their son did not seem to be getting sick as rapidly as the other children had.

Here was a religious couple opposed to abortion who, nevertheless, agreed that in a society where abortion was legal, it made no sense to throw the tissue away when it could save a life. Try arguing with that.

Waxman's next witness was Bernadine Healey of the National Institutes of Health. Health and Human Services was the federal agency which controlled NIH and Dr. Healey reported directly to James Mason, the assistant secretary of health, a passionate supporter of the ban. There was no way Healey could avoid his jurisdiction, short of resigning her post.

Waxman knew that Healey didn't believe in the ban but was forced to defend it. That made her especially vulnerable to reasoned inquiry.

"What was the point in holding detailed inquiries into the ethics of fetal-tissue transplants and then ignoring the advice of the panels—one of which you served on?" Waxman asked.

"Sometimes science has to take a 'time-out' when its goals collide with the moral and ethical concerns of society," Healey replied.

"When can we expect this 'time-out' to be over?" Waxman snapped back.

"I view it as a 'time-out' in the sense that science is an evolving story. As more information comes in, it will be made available to the Department of Health and Human Services. Dr. Mason wants to avoid the one-to-one relationship between abortion and therapy."

On July 25, 1991, the House voted to lift the federal funding ban by 274 votes to 144, only two votes short of the two-thirds needed to override the inevitable veto from President George Bush.

California Parkinson's Foundation, September 1991

The Public Broadcasting System's science show *Nova* had taped George's operation in 1989 and now wanted to document the surgery's effects. Two years after the transplant, the change in George was remarkable. He still had difficulty talking normally, but on a good day he could move around as well as most people. He got up out of chairs without using his hands, walked up and down, did hand-flipping tests with ease. Now that the formal evaluation period was over, George's L-dopa medication had been reduced by two-thirds, from 50 milligrams five times daily before surgery to 30 milligrams three times daily.

If George looked good, Juanita looked great. She had flown down from Montana for the day and walked into the examination room smiling and relaxed. Two years before, Juanita's condition had been unbearable. When off medication, it took her over a minute to get out of a chair; on medication, she rocked back and forth with terrible dyskinesias. Now, with her L-dopa medication reduced by 70 percent, she could function almost normally. She could speak and laugh, stand up and sit down with ease. She spoke warmly of her time in Sweden and told Drs. Widner and Langston that she was now helping around the house with the chores and that she was thinking of going to college.

Widner and Langston were excited but tried to hold back their enthusiasm. The clinical tests they had carried out for two years showed a steady improvement. They were sure that these improvements came from the grafted tissue. But other scientists would be skeptical. To convince them, they needed more evidence. So later that week, George and Juanita were due to fly to Vancouver for a PET scan.

University of British Columbia Medical School, Vancouver

Håkan Widner watched George lying on the table. He was very stiff, as he had taken no L-dopa that morning. In a few minutes a special compound would be given to George intravenously. This compound, [^{18}F] 6L-fluorodopa, would pass into George's brain, where it would be converted into L-dopa and be taken up by the dopamine neurons. Tagged to the fluorodopa molecules would be a source of positrons. By tracking the positrons, the team effectively tracked the way the fluorodopa was absorbed in different regions of George's brain. If everything was done correctly—and this team at the University of British Columbia was one of the best in the business—the scan would yield a very detailed picture of how George's graft was performing.

This was not the first time that George and Juanita had been to Vancouver, nor would it be their last. But on this trip, to save time, Widner had brought them up together. He had asked Juanita if she minded traveling up with George, and she had reluctantly agreed. So the three of them had flown up on Sunday afternoon from California. On the flight up, George had courted Juanita continually, and she had been cool. But once they arrived in Vancouver she had relaxed a little, and she and George had gone off to the smoking room together to chat.

The plan was for George to be scanned on Monday and Juanita on Tuesday. The PET scan technique is extremely complex and many factors have to be controlled for. The first is diet. Since the amino acids from which proteins are made compete with fluorodopa to get into the brain, the scan is greatly affected by what a patient eats just prior to the scan. A huge meal of steak and eggs will produce a different result from a light dish of pasta in the same patient. For this reason, every time George and Juanita came to Vancouver they were given a special low-protein meal.

Second, the team does not want the fluorodopa to be metabolized into dopamine by the body before the fluorodopa gets into the brain. So an hour before the PET scan, George was given carbidopa to block systemic metabolism, so that most if not all of the fluorodopa would go straight to the brain.

Third, as in all imaging, a point of reference is needed. It is not necessary to bolt a stereotactic frame. Instead the patient's head is held steady in a plaster mask. Then the table is moved to position the head in the center of the large detecting ring and a laser registers its position.

Two miles away from where George lay, a cyclotron finished generating a source of positrons. The positrons were attached to fluoride ions which had combined with dopa to make [^{18}F] 6L-fluorodopa. The weakly radioactive material was put into a capsule by the cyclotron technician, inserted into a specially designed pneumatic tube, and fired the two miles to the PET scan center at the University of British Columbia.

Opening the capsule, the PET scan nurse injected about 3 cubic centimeters of the radioactively labeled dopa into George's right arm, and the rest of the team jumped into action. Every thirty seconds for the next five minutes they took blood samples from an arterial line in George's left arm for immediate analysis to determine the level of radioactivity of George's blood. This data was entered into the computer and used to calibrate the instruments.

Then the team settled down to make the scan.

While George lay quietly on the table, a lot was going on beneath his skin. Some of the fluorodopa was converted into dopamine, but because of the carbidopa, most of it made it to his brain. This fluorodopa was converted into dopamine and accumulated by the dopamine uptake system, which now included the grafted fetal neurons. The purpose of the scan was to determine what happened to this dopamine, how and precisely where it was absorbed in George's brain. Over the next two hours, the detectors registered thousands of events and a computer integrated them into an image which showed a picture of how much dopamine was being absorbed by the grafted regions of George's brain—the putamen and caudate.

The results of the PET scans for both George and Juanita were impressive: George's brain cells were retaining 60 percent more dopamine than before surgery, and Juanita's brain was retaining twice as much.

22

Going Public

November 4, 1992

By lunchtime it was clear that the forty-second president of the United States would be Bill Clinton, a pro-choice Democrat. With Clinton in charge, it was quite likely that the funding ban would be overturned and that American women would continue to have the right to choose an abortion. As Bill Langston listened to the early returns on television, he looked over the paper that was soon to be published in *The New England Journal of Medicine.* Håkan Widner was the principal author of the piece and, following established protocol, his name would appear first. Bill Langston, as the senior author, would appear last. In between Widner's and Langston's names would be the names of the other scientists involved: Olle Lindvall, Anders Björklund, Patrik Brundin, Stig Rehncrona, Jim Tetrud, Barry Snow (who supervised the PET scans), and Björn Gustavii (the Lund gynecologist).

After a long review process, *The New England Journal* had decided to publish their work, along with preliminary results from Eugene Redmond's Yale group and Curt Freed's group in Colorado. The three papers coming out together would create quite a stir. On the grapevine, Langston had heard that neither the Yale group nor the Colorado group had observed results as dramatic as they had seen in George and Juanita.

The paper contained the data from George's and Juanita's PET scans, which was very encouraging. Compared with PET scans done before surgery, there was a clear-cut increase in fluorodopa in the areas where the transplants had been done—the caudate and putamen. The results were also sobering. Complex calculations had confirmed that the grafts had managed to innervate only about 10 to 20 percent of George's and Juanita's striatum, and that the new cells seemed to lie in isolated clumps.

If material from up to seven fetuses transplanted bilaterally could only innervate 10 to 20 percent of the caudate and putamen, then it was very doubtful that using a single fetus on one side of the brain—as the other teams had done—could be doing very much. A lot more research would be needed before this technique could be used on a large scale. It was clear to the Swedish team and to Langston that they would have to find some way of improving the yield and also of spreading the grafts out more in the putamen in future operations. The Swedish team were preparing to do their seventh, eighth, and ninth transplant cases. These would be ordinary Parkinson's disease patients who would be grafted bilaterally, like George and Juanita.

Three weeks later, their paper appeared in *The New England Journal of Medicine*. The same week a *Nova* episode aired across America and was seen by millions of people. Whereas the scientific paper described George's and Juanita's improvement in terms of cold objective data, *Nova* showed the partial restoration of two human beings crippled by parkinsonism. And it looked impressive. Before surgery they could not get out of a chair; after surgery they walked around almost normally. Viewers saw George—the original frozen addict—mount a bicycle and pedal away. They saw Juanita smile and joke. It was an amazing transformation.

Together, *The New England Journal* article and the *Nova* program attracted a fair amount of publicity. All the major news services, including the Associated Press, carried stories on fetal-tissue transplants and the surrounding issues. The main television networks covered the story in their prime-time news programs. The local northern California newspapers—the *San Francisco Chronicle* and the *San Jose Mercury News*—did extensive pieces for which Langston was interviewed. Additionally, Langston's and Juanita's picture appeared in *Newsweek* and in a long article about the MPTP cases written for the *Boston Globe*. The Saturday, December 5, 1992, edition of the *New York Times* had an editorial featuring George and Juanita, the couple whose tragic accident in 1982 had precipitated a research revolution. President-elect Clinton vowed that as one of his first acts, he would overturn the funding ban.

It was generally accepted by the scientific community that the Swedish-Californian study eclipsed the other two. The results were more impressive and the science was rigorous.

After watching the *Nova* segment on November 30, 1992, the brother of George's ex-wife felt moved to help him. He went and picked up George from the latest of a dozen halfway houses that George had boarded in and invited him home to live with him and his wife. George Carillo accepted and moved to the suburbs. He started talking about getting a job, earning money, regaining some control over his life, driving a car. In February 1993, George sat the written part of the California driving test and passed as a first stage to taking the driving part of the test.

The next month, George's hallucinations disappeared. Following his postoperative evaluation, his L-dopa dosage had been reduced by two-thirds and now he was benefiting. For the first time in years, George was freed from the terrifying images of snakes and fire. While George was not cured, he had not been so healthy for a decade. He had been given a second chance.

The reports from Montana suggested that Juanita was continuing to improve. She was doing well on low doses of L-dopa (a fifth of what she took prior to surgery), with little in the way of dyskinesias. Even more remarkable was the way she was putting her life back together. A total invalid before the transplant, she was now functioning independently. She had been awarded her own house by the chiefs of the tribe and had enrolled in a junior college, where she had started working on her high school equivalency degree. This required walking some distance each day to class. Her goal was to eventually go into drug counseling.

But the best news was that the Swedish team had agreed to operate on Connie. Even though she hadn't initially been judged suitable for a scientific study, Håkan Widner had managed to convince his colleagues to accept her for a transplant. There were, of course, compassionate grounds for doing the procedure, but Widner argued that she might provide very important data. Connie was in many ways the purest case. Since she had been unmedicated for a number of years, any improvement in her ability to move would incontrovertibly be due to the fetal graft.

There were a few problems. The Swedes would not pay for this operation, and so Langston had to raise some fifty thousand dollars. Also, they would not be rushed into doing it until after they had grafted and done preliminary evaluation on three more British Parkinson's disease patients who were due to come to Lund in 1993. Given the slow and methodical way the Swedes worked, Connie's turn was unlikely to come until 1994.

23

Unending Quest

On September 15, 1993, Langston and his staff moved from their offices in San Jose to a new building in Sunnyvale, California, in the heart of Silicon Valley. Langston took advantage of the move to rename the organization.

The new Parkinson's Institute was the realization of Langston's vision, a comprehensive center for Parkinson's disease and related disorders. For the first time, all of their programs, ranging from day-to-day patient care to basic laboratory research, were now under one roof. Ideas could flow freely between the clinic and the laboratory. An observation in the clinic could be tested in the laboratory within days. New findings in the laboratory could be quickly and efficiently tried out on patients. Staff from all disciplines—clinicians, clinical researchers, and basic scientists—were in constant contact. They ate lunch together, talked about their work over coffee, and rubbed shoulders in the halls.

The clinic has provided care for more than two thousand

patients with Parkinson's disease, and another six hundred with other movement disorders, such as essential tremor, Huntington's chorea, and Tourette's syndrome. With a staff of nearly fifty, including four senior movement-disorders specialists, they were one of the largest programs in the country. Their annual budget for 1994 exceeded 3.5 million dollars.

From a scientific standpoint, Langston no longer expected to find easy answers. He realized that there would be many setbacks before Parkinson's disease was understood and conquered. Langston's experience with deprenyl was a good example, having provided a series of scientific and emotional ups and downs. Jim Tetrud and Bill Langston's pilot study begun in 1987 and involving fifty-four patients had taken three years to complete. The data appeared clear-cut: deprenyl delayed the need for L-dopa in untreated patients by almost one year and, based on clinical evaluations, appeared to slow progression of Parkinson's disease by nearly 50 percent. The much larger DATATOP study of eight hundred patients, published in 1989, had confirmed Langston and Tetrud's smaller study: deprenyl unequivocally delayed the time when a patient required L-dopa therapy, whereas vitamin E (which some doctors thought mopped up free radicals) did not.

Langston had been exhilarated. This was the first time his research had directly benefited patients. But it was not long before clinicians and scientists began to argue that both studies had methodological flaws, and in several instances the attacks had been vicious. One irate physician had written an article in *Neurology,* the most widely read neurology journal, ridiculing the studies.

Tetrud and Langston and the much larger group of investigators involved in the DATATOP study had hypothesized that deprenyl worked by preventing neurons from dying in Parkinson's disease through blocking the production of potentially toxic substances. Such substances were thought to arise from the oxidation of dopamine (thereby producing potentially damaging free radicals) or as the result of exposure to environmental agents such as MPTP, which might be oxidized into toxic compounds. The step that leads to toxicity in both of these situations is mediated by the enzyme MAO B, which deprenyl selectively blocked. It followed that if one or both of these processes was responsible for progressive loss of nerve cells in Parkinson's disease, then blocking MAO B with deprenyl should slow or halt disease progression.

But some critics argued there was an alternative explanation for deprenyl's effects that had nothing to do with protecting neurons: deprenyl was acting symptomatically, producing a transient day-to-day "L-dopa-type" effect in the patients. Such a symptomatic effect could mimic slowing of disease progression since, on a day-to-day basis, patients would look like they were better off than the control group not taking deprenyl. How could scientists decide which explanation was true?

Because it is currently impossible to measure the rate at which neurons degenerate in the living human brain, the deprenyl studies had to infer neurological degeneration from the patient's clinical condition. The designers of the DATATOP study had realized, however, that clinical examination by itself cannot directly differentiate whether or not deprenyl is protecting neurons or merely pharmacologically alleviating symptoms. They came up with a protocol to address this problem by building into the trial two ways of monitoring for a symptomatic effect: one at the beginning of the trial (the "wash-in" phase) and one at the end (the "washout" phase).

Shortly after the study started, both the placebo and deprenyl groups were carefully evaluated to see if they had improved. An immediate improvement in the deprenyl group but not the placebo group would be evidence of deprenyl acting symptomatically. This protocol was referred to as the "wash-in," and it did show that deprenyl had a minimal symptomatic effect, an effect so small that it had not shown up in the fifty-four patients studied by Tetrud and Langston. It could, however, be seen "statistically" in the much larger study of eight hundred patients. It was this evidence that the critics pointed to when declaring deprenyl was symptomatic and not neuroprotective. But to Langston, it seemed difficult to believe that such a small benefit could delay the need for L-dopa for nearly a year.

More important in Langston's opinion was the "washout" part of the study, when treatment (either deprenyl, vitamin E, or placebo) was stopped for an entire month, after which the patient's disease state was evaluated. This was to give the drug time to wash out. If deprenyl's effect was symptomatic on a day-to-day basis, one would expect these patients to get worse immediately following cessation of the drug. But in general they didn't.[1]

In any case Langston believed that regardless of whether deprenyl lived up to early hopes, it had ignited great interest in the idea of neuroprotective therapy—finding ways to protect neurons from dying as the result of a neurodegenerative disease. And in 1993, a colleague and good friend of Langston's, Dr. Warren Olanow, who had recently become chairman of the department of neurology at Mt. Sinai Medical School in New York, reported a new study of deprenyl that addressed some of the criticisms of the earlier studies. The results clearly supported the neuroprotective hypothesis rather than the view that deprenyl was merely therapeutic. Another investigator in Canada, Dr. William Tatton, had discovered an entirely different mechanism of action of deprenyl, and was suggesting that it could "rescue" damaged or dying neurons when given in lower doses than those used in the two original deprenyl studies.

Deprenyl was not the only drug Langston and his colleagues were interested in. The Parkinson's Institute was now conducting more than a dozen new drug trials for Par-

[1]In a later modification of the study, when patients were washed out before reaching the endpoint of the study and for a longer period of time (i.e., more than a month), there was a degree of deterioration but, at the final evaluation, the deprenyl patients were still better off than the placebo group.

kinson's disease, including trials of new MAO inhibitors, new dopamine agonists (drugs that are not converted to dopamine once they reach the brain, but rather stimulate dopamine receptors directly), and a new drug that blocks catochol-O-methyltransferase, another enzyme that breaks down dopamine. In addition, in the laboratory they were testing new antioxidants (free-radical scavengers) that held some real promise for Parkinson's disease. This work was being done in collaboration with a start-up biotechnology company called Centaur Pharmaceuticals, located just a few miles away.

Such drug trials have many benefits. Patients gain access to new drugs years before they reach the marketplace. Medical researchers learn more about the effects of these drugs in patients. And the trials are usually fully funded by the drug companies, which means not only that patients don't have to pay for treatment but also that the studies help support the institute. But there are drawbacks to these clinical trials. Since the drugs are experimental, there is always the possibility of unexpected side effects. Also, many of these studies are "controlled," meaning that a portion of the patients will be receiving a placebo rather than an active drug; neither the patient nor the physician knows which is being taken (this is referred to as a "double-blind" study). Not surprisingly, many patients are unwilling to enter into such studies.

If clinical research was thriving, basic research into the cellular mechanism of cellular degeneration in Parkinson's was also flourishing both at the institute and elsewhere. Following the discovery that MPP+ blocks a critical step in mitochondrial respiration, scientists had hypothesized that mitochondria might hold one of the keys to neurodegenerative diseases like Parkinson's. Mitochondria are small organelles that are found by the hundreds in all living cells in the body. They are like miniature energy plants that supply power for all of the cell's activities. Destroy enough mitochondria, and it's like turning off the heat and electricity of the cell: it loses power and dies. Inspired by the basic research on MPTP, investigators examined a variety of tissues from patients with Parkinson's disease and found abnormalities in mitochondrial respiration that were very similar to those produced by MPP+, the toxic metabolite of MPTP. This discovery opened up a new line of research which quickly broadened to the entire question of brain energy metabolism and other human neurodegenerative diseases as well.

But there were many other theories on why dopaminergic brain cells die in Parkinson's disease. The discovery of MPTP had not only led to an effective animal model and inspired new therapeutic strategies; it also raised the possibility of an environmental cause for Parkinson's disease. Since Calne and Langston's article in *The Lancet* highlighting the environmental hypothesis for Parkinson's disease, researchers had been searching actively for chemical substances that might account for the incidence and distribution of the disease. Scientists had investigated lifestyle, location, occupation, groundwater, pesticides, herbicides, paper and pulp mills, and many other factors. But so far, the evidence did not point to a single cause, and what evidence existed was controversial.

Langston had come to have enormous respect and sympathy for scientists involved in epidemiological research. Being an epidemiologist is in many ways a thankless task. The studies take a very long time; the research for a single paper can take decades. And because there are so many potential pitfalls in epidemiological research, when studies are submitted for funding (or publication when complete), they can easily become the center of controversy and ruthless criticism, often fueled by economic and political objectives. The seminal studies on smoking and lung cancer done by Sir Richard Doll in the 1950s are still attacked by scientists funded by the tobacco industry.

Despite the difficulties, Langston felt that such studies would help point the way toward the cause of Parkinson's disease if enough care was taken. In 1986, with a grant from the Centers for Disease Control, he had set up the Parkinson's Epidemiology Research Committee (PERC), an eclectic group of experts from a variety of disciplines including biostatistics, toxicology, epidemiology, neurology, and environmental health. PERC was charged with the development of epidemiologic protocols for identifying industrial, agricultural, and naturally occurring environmental chemicals and compounds, including natural products and plants, that might have neurotoxic effects similar to MPTP. They were to assess past and current epidemiological efforts, and to consider future methods for exploring the environmental agents that might play a role in Parkinson's disease.

As they examined dozens of studies that had sought a relationship between prevalence of Parkinson's disease and, for example, rural living (an observation first made by Dr. Ali Rajput in Saskatchewan Province in Canada in 1984), pesticide exposure, proximity to wood pulp mills, or drinking well water, definite conclusions were evasive. Some studies reported a higher incidence of Parkinson's in areas with pesticides and farming; some didn't. While generally it seemed that Parkinson's disease was more common in intensively farmed rural areas, no exogenous chemical had been identified which caused irreversible and progressive damage to the nigrostriatal pathway, similar to the damage found in Parkinson's disease.

So either the environmental cause of Parkinson's disease is still out there waiting for someone to find it, or the environment is only part of the story. The other part may well be genetic susceptibility.

Back to Genetics

Following a series of twin studies in the mid-1980s, most scientists abandoned a genetic cause for Parkinson's disease. In the 1983 NIH study on twins, for example, out of forty-three twin pairs in which at least one twin had Parkinson's disease, there was only one pair where the matching twin had Parkinson's disease as well. This rate was no higher than could be found in the general population. But since that time, a few scientists had come to doubt the study, including its principal author, Roger Duvoisin. By the late 1980s, Duvoisin had begun to suspect that the diagnostic screening of the cohort was flawed and that the degree of concordance was much higher than reported.

Several examples of "familial parkinsonism" had come to light in which Parkinson's disease seemed to be inherited in a Mendelian fashion. In one Italian-American family, more than forty-five members over four generations were found to be affected by a Parkinson's disease-like disorder which appeared to be inherited in an autosomal dominant manner (50 percent of the offspring of an affected individual are affected). There was little doubt about the diagnoses in this family (a so-called kindred) since autopsy verification of diagnosis was available from at least two generations.

While many investigators do not believe that the majority of Parkinson's disease is inherited in this way, Duvoisin and others were now saying that it is. One of the observations cited in support of this position is that historically many of the cases are missed because the disease has a long preclinical phase. Many researchers believe that for years, perhaps decades, before clinical symptoms appear, degeneration of dopaminergic cells has already started. Approximately 80 percent of striatal dopamine must be lost before the symptoms of PD become evident. Researchers studying twins cannot look into their brains to see what fraction of the dopamine-producing cells in the substantia nigra are dead. They cannot detect Parkinson's disease clinically before there are any clinical symptoms. But what if they had a way of estimating the amount of dopamine in the brain?

There is a way, albeit an expensive one—the PET scan. Langston's group of asymptomatic MPTP drug abusers had been the first patients to be PET scanned for this purpose, back in 1985. Langston had continued to follow this group. In PET scans carried out in 1992, four of these patients showed a clear-cut decline in their striatal dopamine uptake from their 1985 PET scans, indicating progression toward clinical parkinsonism. In fact, the father of one of the original frozen addicts had taken MPTP in the summer of 1982 with his son and in the early 1990s had developed full-blown clinical parkinsonism. Langston felt that this case could well be pivotal in convincing other scientists that exposure to an environmental toxin can show up years later as a progressive neurodegenerative disease.

By 1993, many other scientists were now using PET scans for research. David Brooks at London's Hammersmith Hospital carried out PET scans of clinically asymptomatic individuals who might be progressing toward parkinsonism. He scanned asymptomatic individuals in Parkinson's disease kindreds (like the Italian families), and a cohort of identical twins. He found many cases in which individuals with no recognizable symptoms showed early signs of degeneration as evidenced by a decreased uptake of fluorodopa in the striatum as compared with normal controls.

Most remarkable was the data on twins. Ten pairs of identical twins were given a PET scan. In each pair, only one twin showed clinical symptoms of Parkinson's disease at the time. However, the PET scans revealed that, in addition to the ten clinically affected twins, four of the unaffected twins showed decreased uptake of flurodopa in the striatum as well. In light of this new evidence, Duvoisin and others were now arguing that the 1983 and 1984 twin studies which found no concordance (as measured clinically) would have to be reinterpreted.

In 1994, the majority of investigators believe that the cause of Parkinson's disease may well be due to a combination of environmental factors and genetic susceptibility. Teasing out the relative contributions of each is not likely to be easy. But thanks to a lucky discovery, Langston and his colleagues are pursuing promising new research avenues that could answer these questions. While compiling a list of existing databases for PERC, Dr. Jonas Ellenberg, chief of the Biometry and Field Studies Branch of NINDS, stumbled on a little-known database containing fifty years of records on thousands of identical twins.

In 1946, surgeon Michael DeBakey, then working in the army surgeon general's office, had campaigned to ensure that the medical community would have access to World War II medical records. His campaign was successful, and the National Academy of Sciences created a follow-up agency to keep tabs on twins, since they seemed to be the most interesting population of veterans for medical studies. Among the millions of young men who served in the war were over 20,000 white twins—both identical and fraternal—and during the 1950s their records were gathered into a huge registry.

The National Academy made little attempt to publicize this registry, and knowledge of its existence declined. But when the PERC researchers heard about it, they realized that this population represented a research gold mine for the study of age-related neurodegenerative diseases. Many of the twins, in their late teens or twenties when the war started, are now in their sixties and seventies, approaching the onset age for Parkinson's disease. This database should have the information to settle some of the questions on the etiology of Parkinson's disease. Langston's colleague at the Parkinson's Institute, neuroepidemiologist and movement disorders specialist Caroline Tanner, is heading a multimillion-dollar collaborative twin study to probe the relative importance of genetic and environmental factors in Parkinson's disease. At the moment, there is simply no better way to explore these factors than in this unique population.

Neurotransplantation in the United States

On Tuesday, January 5, 1994, the federal government approved the first federal grant to study the effects of implanting fetal tissue into the brains of Parkinson's disease patients. The $4.5 million grant was earmarked for Dr. Curt

Freed of the University of Colorado Health Sciences Center in Denver to carry out fetal-tissue transplants in twenty patients. Seven years after Ed Oldfield had asked Irwin Kopin for permission to do this procedure, fetal-tissue research was once again part of the mainstream of publicly funded biomedical research.

Langston expected the field of neurotransplantation to develop steadily. Much of the refinement of the surgical procedure was being carried out in primates with MPTP-induced parkinsonism because they represented such a faithful model of the human condition. There were still many questions yet to be answered. What is the best type of instrument to inject these donor cells with? Where precisely should the grafts be placed in the caudate and putamen for the best results? How many injections are desirable? What parkinsonian symptoms respond the best to fetal implants? Should immunosuppression be used to prevent the small possibility of graft rejection? How can the percentage of surviving cells, now only about 10 percent, be increased?

While Langston planned to start a transplantation program in California, building on the unique experience of his colleagues at Lund, the institute was pursuing other promising techniques as well. It now housed the preclinical research program for Somatix Therapy Corporation, a group working on the development of genetically engineered cell lines that produced dopamine. This new approach involved taking the subject's own skin cells and inserting genes into them so that they made dopamine. These "born again" skin cells were then transplanted into the brain where they replace the missing dopamine. Theoretically, there were many advantages to such an approach. Being the patient's own tissue, the cells would not be rejected. Such cell lines would avoid the ethical controversy and the practical difficulties that attended the procurement of fetal tissue.

The first step was to try genetically engineered cell lines in primates. If the technique could work in monkeys with MPTP-induced parkinsonism, it might well work in humans with ordinary Parkinson's disease. Somatix was also attempting engineering cells to make the growth factors that might revitalize malfunctioning neurons and possibly reverse parkinsonism.

There was another promising area of research that had been reopened by a scientist at Emory University, Dr. Malon DeLong. Using the MPTP research model, DeLong had made great progress unraveling the circuitry of the basal ganglia. In 1989, he reported in *Science* that a small area in the brain known as the subthalamic nucleus was overactive in monkeys with MPTP-induced parkinsonism, and that lesioning this nucleus, which connected to the internal portion of the globus pallidus, abolished the parkinsonism. This was reminiscent of a procedure tried on humans more than four decades ago called pallidotomy, in which a tiny portion of the globus pallidus (another part of the basal ganglia) was lesioned. Results of pallidotomy had been very mixed. When successful, it appeared to immediately reverse all of the major symptoms of Parkinson's disease. But success was inconsistent as surgeons had no reliable way of locating the critical area of brain to lesion. DeLong's work had dramatically revived interest in this technique and many scientists, Langston included, thought that research into pallidotomy was now very important.

Despite all these developments, Langston would never forget what science owed to George, Juanita, Connie, David, Bill, and Toby. Twelve years ago when he had been called out of the general neurology clinic to see George Carillo, Langston had been annoyed at the inconvenience. Yet George and the other MPTP cases had become the center of his life and work. They had turned him from a struggling clinical researcher at an obscure county hospital into a focused neuroscientist working at a world-class scientific institute which he had been instrumental in founding. They had changed not only his prospects but also those of Parkinson's disease research.

Two of the MPTP cases—George and Juanita—had in

turn benefited from that research. They had made a partial recovery following surgery and now had the prospect of living a more normal life.

Three of them—David and Bill Silvey and Toby Govea—would probably never be accepted for experimental fetal-cell surgery, unless they were able to avoid future run-ins with the law.

The remaining MPTP index case was Connie. Since her dosage of L-dopa had been cut back radically in 1987 and discontinued completely in 1990 because of terrifying hallucinations, Connie had not been able to move her hands, arms, or legs. She could not move her head; she could not move her chin. Langston had no way of knowing whether or not she was still mentally normal inside her frozen body. He was not sure how anyone could stay sane for four years understanding everything the people around her are saying, but unable to communicate her wishes and desires to those people.

Even highly disabled people such as those with cerebral palsy are able to communicate by flicking an eyelash, or sticking out a tongue. Modern computer technology had been able to build on these simple voluntary movements to enable such people to initiate communication. But these sophisticated technologies did not work for Connie, because Connie had virtually no voluntary control over any of her muscles. In 1993, Langston had raised the money to fly out a team of communication experts from the Center for Applied Special Technologies (CAST) in Peabody, Massachusetts. But the CAST team, who had designed electronic solutions enabling some of the most severely disabled people to communicate, were unable to help Connie. They could find no "access point" through which Connie could express her intentions. Wherever they placed their tiny switches—above her eyes, on her hand, under her chin—Connie was unable to move her muscles to activate the switch.

However, for two reasons Langston suspected that Connie was still engaged by the outer world. One was humor. A good joke would cause Connie to smile. As smiling is an involuntary response, Connie didn't have to do anything to make it happen. It just broke out. This was enormously important. Humor is very complex, and the ability to laugh in the right places indicated that Connie was cognitively alert. The other ability Connie had was some faint vocalization. She could with great difficulty make a soft moaning sound and with practice people could learn to distinguish yes and no.

On Monday, April 4, 1994, with funds raised by a local columnist, Jim Trotter, who had written a series of articles on her plight, Connie left for Lund, Sweden, for the operation that held out her only hope of a better future. Connie's family came to see her off at San Francisco airport. Because Connie's mother, Nellie, was too frightened to fly, Trina, a close family friend, volunteered to accompany Connie on the long trip and stay with her at Lund during the surgery. Langston watched as Connie, Trina, and Håkan Widner boarded a United flight to Los Angeles for the first leg of their journey. In LA they would transfer to an SAS plane bound for Copenhagen. Then they would take the Hovercraft to Malmö, arriving in Sweden at 3:35 P.M. on Tuesday, April 5.

This moment had been a long time coming. It was because of Connie that Langston had begun his association with the Swedish transplant program. Now she was finally getting her chance. The Swedes had promised to take Connie if they achieved reasonable success with George and Juanita, and they had kept their word. But Connie was more than just a desperately sick patient to the Swedish team. Increasingly, they had come to see her case as scientifically important. Connie—the worst-affected of the MPTP patients—was a benchmark test for neurotransplantation. A significant improvement in Connie's state following surgery would dramatically demonstrate the efficacy of fetal-tissue transplants uncomplicated by medication. If a transplant could help her, it should in principle be able to help even the most advanced Parkinson's disease patient.

The days before surgery were traumatic. Nine thousand

miles from home, Connie lost her nerve and decided against having the operation. After Trina placed a few frantic telephone calls to California—to Connie's family in Greenfield and to Langston in San Jose—she changed her mind again and consented to the surgery. The next day, Thursday, April 7, Connie's temperature soared. Over the next twenty-four hours, Connie's physicians contacted infectious disease specialists throughout Sweden for advice. It looked certain that the operation would have to be postponed. Then the fever disappeared as suddenly and mysteriously as it had appeared.

A day later, on Friday, April 8, Stig Rehncrona implanted material from 4.5 fetuses into Connie's left-side putamen. Two weeks later he repeated the procedure on the right side of her brain. The operations went well and Connie returned home to California on the evening of May 2.

Over the summer, Connie's family reported that she seemed to be getting better. Connie was sleeping better, was less depressed, and seemed to have regained some of her lost function.

On September 15, 1994, five months after the fetal-tissue surgery, Connie's family drove her up to the Parkinson's Institute for a full examination. When Langston entered the examining room, he could scarcely believe his eyes. Dressed in a striking blue sweater and slacks, Connie was bright and alert, and actually wearing make-up. She looked pretty. She smiled. A certain amount of expression had returned to her face. Her speech was easier to understand. When Langston asked her a question she could communicate the answer, but as yet she was unable to initiate a dialogue. While still incapacitated, the improvement in just five months was astonishing. Nothing like this had happened with George or Juanita.

Langston began to examine Connie in detail. Much of her muscle stiffness was now gone. Previously she could hardly move her arms at all. Now she rapidly raised and lowered her arms with almost normal speed. The movements were mechanical, lacking fluidity—almost as if her arm was being pulled up and down with a string like a marionette—but the fact that she could move her arms at all was remarkable. Next Langston lifted Connie to a standing position and discovered she could walk with minimal assistance, bearing almost all of her own weight.

Langston was deeply moved. Here was a woman he had known for twelve years, yet never really known at all. Here was the woman who had inspired his collaboration with a team of Swedish scientists half a world away. There had been many times he was afraid she might die. There had been times when he had felt close to tears because he could offer her so little to ease her suffering and isolation. Today he felt a mixture of hope and fear. For the first time he allowed himself to believe that the Swedes had restored at least some of what MPTP had taken away. Yet Langston also feared the future. If for any reason the transplant failed, how could Connie deal with being plunged back into isolation? Don't get too excited, Langston told himself. Even more important, don't let *her* get too excited. Fetal-tissue transplants take time. It would be a year or more before they could be sure how well the transplant had worked. There would be dozens of clinical examinations and several trips to Vancouver for PET scans.

But now, Langston wanted to ask Connie many questions. Questions about what it had been like to be imprisoned in her body; about what she missed most; about how she kept her sanity for so many years. Yet she was still so fragile and it required an enormous effort on her part to communicate. He decided to wait until she was stronger.

Langston smiled and looked into Connie's eyes, which seemed to be smiling back at him. He took Connie's hands gently in his own and whispered, "Welcome back."

Epilogue

On November 29, 1994, the day after his 42nd birthday, Bill Silvey died. He had been in very poor health for months with a chronic intestinal disorder, and his weight had dropped to less than 100 pounds. Realizing the scientific importance of Bill's case, his family gave permission for his brain to be examined for research purposes.

When Dr. Lysia Forno examined sections of the upper brainstem under the microscope, she found, as expected, that the substantia nigra was almost completely gone, destroyed by MPTP. She also noted subtle evidence that nerve cells in the area were continuing to die, even though the MPTP exposure had been over twelve years earlier. This observation, if confirmed in other cases, could have profound implications, indicating that a brief exposure to an environmental toxin might set in motion a slowly progressive process that could continue for years. Might such a process underlie neurodegenerative diseases such as Parkinson's or Alzheimer's?

Even in death, Bill Silvey was helping scientists unravel the secrets of the brain in health and disease.

Acknowledgments

Many people contributed time and assistance in the preparation of this manuscript. We should like to acknowledge the help and advice of the Lund transplant team, especially Olle Lindvall, Anders Björklund, and Patrik Brundin. We owe a special debt of thanks to Håkan Widner, who read the manuscript at an early stage and made many suggestions. Our thanks also go to the staff at the Parkinson's Institute, notably Ian Irwin, Jim Tetrud, David Rosner, Mary Lee, and Caroline Tanner. We acknowledge Sandy Markey and Irv Kopin of NIH and Neal Catagnoli of UC San Francisco for checking particular sections. We are grateful also to the Santa Clara County police and crime lab, especially Dave Weidler, Jim Norris, and Halle Weingarten.

Having dreamt for some years of writing this book, we should like to thank the people who helped make it happen, notably our agent, Jill Kneerim, and our editor, Linda Healey, whose help and support were invaluable. Of our families, who endured our absences during the writing of this book, we ask forgiveness.

Our greatest debt of thanks is to the people directly affected by this tragedy, who, despite their burden, have been unfailingly forthcoming and helpful in our continuing attempts to uncover the mysteries of parkinsonism and to communicate neuroscience research to a wider audience. Our thanks and hopes go to the Sainz, Silvey, Govea, and Lopez families and to Connie, George, Juanita, David, Bill, and Toby.

Index